Variational Methods in
Electron-Atom
Scattering Theory

PHYSICS OF ATOMS AND MOLECULES

Series Editors:

P.G. Burke, *Queen's University of Belfast, Northern Ireland*
and
H. Kleinpoppen, *Institute of Atomic Physics, University of Stirling, Scotland*

Variational Methods in
Electron-Atom Scattering Theory

Robert K. Nesbet

IBM Research Laboratory
San Jose, California

PLENUM PRESS · NEW YORK AND LONDON

√ 6414-8051
PHYSICS

ISBN 0-306-40413-3

To Helen, Anne, Susan, and Barbara

Preface

The investigation of scattering phenomena is a major theme of modern physics. A scattered particle provides a dynamical probe of the target system. The practical problem of interest here is the scattering of a low-energy electron by an N-electron atom. It has been difficult in this area of study to achieve theoretical results that are even qualitatively correct, yet quantitative accuracy is often needed as an adjunct to experiment.

The present book describes a quantitative theoretical method, or class of methods, that has been applied effectively to this problem. Quantum mechanical theory relevant to the scattering of an electron by an N-electron atom, which may gain or lose energy in the process, is summarized in Chapter 1. The variational theory itself is presented in Chapter 2, both as currently used and in forms that may facilitate future applications. The theory of multichannel resonance and threshold effects, which provide a rich structure to observed electron–atom scattering data, is presented in Chapter 3. Practical details of the computational implementation of the variational theory are given in Chapter 4. Chapters 5 and 6 summarize recent applications of the variational theory to problems of experimental interest, with many examples of the successful interpretation of complex structural features observed in scattering experiments, and of the quantitative prediction of details of electron–atom scattering phenomena.

The author is indebted to many colleagues in both the theoretical and experimental fields for communicating their work in advance of publication and for granting permission to reprint figures, as cited in the text. He wishes to thank J. D. Lyons, R. S. Oberoi, A. L. Sinfailam, and L. D. Thomas for their collaboration in obtaining many of the results presented here. A special word of thanks is due to Blanca Gallegos, who prepared a long and difficult manuscript with remarkable speed and accuracy.

Robert K. Nesbet

Contents

4. Computational Technique

5. Applications to One-Electron Atoms

6. Applications to Other Atoms

1

Quantum Mechanics of
Electron – Atom Scattering

Introduction

Collisions between free electrons and atoms, molecules, and their ions
are a driving mechanism for many processes of great practical importance
and scientific interest. In complex excited media, ranging from stellar
atmospheres to laser systems, these processes can be analyzed and understood only if the various competing subprocesses are correctly characterized. This book is concerned with quantitative theoretical methods that
have helped to elucidate one class of such subprocesses—scattering of
electrons by neutral atoms at energies below the ionization threshold.

In this area, simplifications due to high incident electron velocity or to
the dominant scattering effect of a Coulomb field are absent. Typically, the
incident electron has energy comparable to valence electrons of an N-
electron target atom, and an $(N+1)$-electron problem must be solved.
Low-energy scattering by a neutral atom is dominated by the long-range
polarization potential that results from dynamical distortion of the atom by
the incident electron. At short range, the external electron becomes part of a
transient negative ion whose short-lived quantum states produce resonance
structures in scattering cross sections. Energy can be transferred between
the incident electron and the target atom, inducing transitions from the
initial atomic state. As the scattering energy is increased, successively higher
excited states of the atom become energetically accessible. Characteristic
energy-dependent scattering structures occur at each excitation threshold.

Theory relevant to these aspects of electron scattering by complex
atoms will be emphasized. Sufficient progress has been made with this

1

problem, using the variational methods that will be presented here, that a detailed exposition and survey of results appear to be justified.

Much less progress has been made with electron–molecule scattering theory. Even for fixed nuclear positions, the reduced spatial symmetry of molecules makes theoretical calculations for molecules more complex than for atoms. Electron impact excitation of vibrational or rotational motion greatly complicates the theory of electron–molecule scattering. Applications to molecules are beyond the scope of the present work. Nevertheless, much of the formal analysis presented here is relevant to molecules in the approximation of fixed nuclear positions.

This chapter will present a survey of the quantum mechanics of electron–atom scattering, applicable to low-energy scattering by complex atoms. A thorough treatment of the formal quantum theory of scattering has been given by Wu and Ohmura (1962), by Newton (1966), and more recently by Joachain (1975). Mott and Massey (1965) survey early developments and applications of electron–atom scattering theory. Geltman (1969) and Brandsden (1970) cover basic theory relevant to electron–atom collisions. It will be assumed that the reader is familiar with potential scattering theory, of which Burke (1977) gives a concise review.

1.1. Structure of the Wave Function

Scattering of an electron by an N-electron atom can be described by a stationary state Schrödinger wave function of the form

$$\Psi_s = \sum_p \mathscr{A}\Theta_p \psi_{ps} + \sum_\mu \Phi_\mu c_{\mu s}. \tag{1.1}$$

The index s denotes a particular degenerate solution at given total energy E. Here Θ_p is a normalized N-electron target state wave function and ψ_p is a one-electron *channel orbital* wave function antisymmetrized into Θ_p by the operator $\mathscr{A}$. The functions Φ_μ are an orthonormal set of $(N + 1)$-electron Slater determinants constructed from quadratically integrable orbital (one electron) functions. The channel orbital ψ_{ps} is characterized by orbital angular momentum l_p and asymptotic energy $E - E_p$, where E_p is the energy of state Θ_p. This energy is positive for *open* scattering channels, defining a wave number or electron momentum (in atomic units) k_p such that

$$\tfrac{1}{2}k_p^2 = E - E_p. \tag{1.2}$$

If $E - E_p$ is negative, the channel is closed. Then k_p is replaced by $i\kappa_p$ with $\kappa_p > 0$.

For low-energy scattering by light atoms, the wave function ψ can be taken to be an eigenfunction of $\mathbf{L}^2$, $\mathbf{S}^2$ and of parity π. Then instead of simple

Slater determinants, the functions Φ_μ can be taken to be antisymmetrized *LS* eigenfunctions. Since these in turn can be expressed as linear combinations of Slater determinants, the expansion indicated in Eq. (1.1) is completely general. To simplify practical calculations, *LS* eigenfunctions are used, and the antisymmetrizing operator $\mathscr{A}$ in Eq. (1.1) is extended in definition to include angular momentum coupling. For heavy atoms, *jj* coupling can be more appropriate, with a corresponding interpretation of the individual terms in Eq. (1.1).

Channel orbital functions are of the form shown in Eq. (1.3), including both angular and spin factors,

$$\psi_{ps} = r^{-1}f_{ps}(r)Y_{lm_l}(\theta, \phi)u_{m_s}, \tag{1.3}$$

where $f_{ps}(r)$ satisfies the usual boundary conditions at $r = 0$. It is assumed the ψ_{ps} is orthogonal (by construction) to all orbital functions used in constructing functions Θ_p and Φ_μ. For an open channel, the asymptotic form of f_{ps} for large r is governed by a phase shift η_p such that

$$f_{ps}(r) \sim k_p^{-1/2} \sin (k_p r - \tfrac{1}{2}l_p\pi + \eta_p), \tag{1.4}$$

for single-channel scattering by a neutral atom. For Coulomb or dipole scattering this functional form must be suitably modified. The normalization corresponds to unit flux density for a free electron. For multichannel scattering

$$f_{ps}(r) \sim k_p^{-1/2}[\sin (k_p r - \tfrac{1}{2}l_p\pi)\alpha_{0ps} + \cos (k_p r - \tfrac{1}{2}l_p\pi)\alpha_{1ps}]. \tag{1.5}$$

The coefficients α_{ips}, with $i = 0$ or 1, determine scattering matrices and cross sections.

Closed-channel radial functions $f_{ps}(r)$ must satisfy the same boundary conditions at $r = 0$ and the same orthogonality conditions as open-channel functions, but the closed-channel radial functions are quadratically integrable. They vanish as $r \to \infty$ in a way determined by the detailed solution of the Schrödinger equation. It should be noted that the term in $\exp(-\kappa_p r)$ arising from analytic continuation of Eq. (1.5) below a threshold at which k_p vanishes is dominated asymptotically by terms in reciprocal powers of r due to long-range interchannel multipole potentials. In electron–atom scattering, such terms define the asymptotic behavior of closed-channel orbitals.

The orthogonality conditions imposed on $f_{ps}(r)$ ensure that each term in the first summation in Eq. (1.1) is orthogonal to the Hilbert space of functions $\{\Phi_\mu\}$ used to construct the quadratically integrable function

$$\Psi_Q = \sum_\mu \Phi_\mu c_{\mu s}. \tag{1.6}$$

The coefficients $c_{\mu s}$ are determined variationally. Slater determinants Φ_μ can be defined in terms of virtual excitations of an N-electron reference determinant Φ_0, itself defined as an antisymmetrized product of N orthonormal *occupied* orbital functions $\phi_i, \phi_j, \ldots$. Virtual excitations are defined by replacing some n specified occupied orbitals of Φ_0 by $n + 1$ one-electron functions drawn from a set of *unoccupied* orbitals $\phi_a, \phi_b, \ldots$ that are mutually orthonormal and orthogonal to the occupied set. These orbitals are all quadratically integrable functions of the space and spin variables of a single electron. An assumed denumerable set of orbitals $\{\phi_i; \phi_a\}$ generates a uniquely defined basis $\{\Phi_\mu\}$ for the $(N + 1)$-electron Hilbert space. A typical Slater determinant Φ_μ can be denoted by

$$\Phi_{ij\ldots}^{abc\cdots}, \qquad i < j < \cdots \leq N < a < b < c < \cdots \qquad (1.7)$$

where the notation implies that $(\phi_a, \phi_b, \phi_c, \ldots)$ replace $(\phi_i, \phi_j, \ldots)$ in reference determinant Φ_0 in the order specified. Appropriate normalization of Φ_μ is implied.

An oscillatory function with nonvanishing asymptotic amplitude cannot be represented as a finite superposition of quadratically integrable functions. For this reason, the open-channel terms $\mathscr{A}\Theta_p\psi_{ps}$ in Eq. (1.1) remain distinct from the Hilbert space component for any calculation using a finite orbital basis. In contrast, closed-channel terms can be included either explicitly or in the Hilbert space component Ψ_Q. In *close-coupling* theory, part of Ψ_Q is replaced by the form

$$\sum_\gamma \mathscr{A}\Theta_\gamma\psi_{\gamma s}. \qquad (1.8)$$

This form is analogous to the open-channel part of Eq. (1.1), except that the functions Θ_γ represent target atom states with E_γ greater than E, corresponding to closed channels. The coupled function $\mathscr{A}\Theta_\gamma\psi_{\gamma s}$ can also be represented, as in the final term of Eq. (1.1), as a linear combination of the quadratically integrable functions Φ_μ.

The target state wave functions $\{\Theta_\gamma\}$ need not correspond to specific stationary states, although they should be orthogonal to the open-channel states $\{\Theta_p\}$. In close-coupling theory, polarization effects can be represented by using *pseudostates* $\Theta_{\gamma(p)}$ that correspond to the first-order perturbation of Θ_p by a polarizing field. Pseudostates and the resulting polarization potentials will be discussed in Section 1.4.

An implicit variational solution for the coefficients $c_{\mu s}$ in Eq. (1.1) can be incorporated into a modified Schrödinger equation for the channel functions $\mathscr{A}\Theta_p\psi_p$, including either open or closed channels, by a partitioning technique used in the resonance theory of Feshbach (1958, 1962). If a projection operator Q is defined by

$$\Psi_Q = Q\Psi_s = \sum_\mu \Phi_\mu(\Phi_\mu|\Psi_s) \qquad (1.9)$$

and if operator P is the orthogonal complement of Q, then

$$\Psi_P = P\Psi_s \cong \sum_p \mathscr{A}\Theta_p \Psi_{ps}. \qquad (1.10)$$

The modified Schrödinger equation, expressed in terms of the $(N + 1)$-electron Hamiltonian operator H, with

$$M = H - E, \qquad (1.11)$$

is

$$M'_{PP}\Psi_P = \{M_{PP} - M_{PQ}(M_{QQ})^{-1}M_{QP}\}\Psi_P = 0, \qquad (1.12)$$

obtained by solving the Schrödinger equation formally for Ψ_Q in terms of Ψ_P then substituting into the equation for Ψ_P. The operator M_{QQ}^{-1} is an $(N + 1)$-electron linear integral operator whose kernel is

$$\sum_\mu \sum_\nu \Phi_\mu(H - E)_{\mu\nu}^{-1}\Phi_\nu^*. \qquad (1.13)$$

Equation (1.12) provides a common basis for the major computational methods used in low-energy electron-scattering theory.

Effective one-electron equations can be derived as matrix components of Eq. (1.12) with respect to target states Θ_p by integrating over angular momentum and spin factors of the channel orbitals Ψ_p to obtain coupled equations for their radial factors $f_{ps}(r)$. When appropriate normalizing factors are included,

$$(\Theta_p|\Psi_P) = \psi_{ps}. \qquad (1.14)$$

The matrix operator acting on channel orbitals is

$$\tilde{m}^{pq} = (\Theta_p|M'_{PP}|\mathscr{A}\Theta_q). \qquad (1.15)$$

The projection of $\tilde{m}^{pq}$ onto orbital angular and spin eigenstates defines a radial operator m^{pq} that acts on $f_{qs}(r)$. The terms in M'_{PP} containing $(M_{QQ})^{-1}$ define a matrix _optical potential_ that acts on the channel orbitals. This operator describes correlation and polarization effects.

When several target states included in Eq. (1.1) have the same total quantum numbers ($LS\pi$ in LS coupling), it will be assumed that the N-electron Hamiltonian matrix has been diagonalized over these states. Then the energy values E_p are eigenvalues of this matrix. The functions $\{\Theta_p\}$ are assumed to be orthonormal.

1.2. Cross Sections

If there are n_C open channels at given total energy E, there are n_C linearly independent degenerate solutions of the Schrödinger equation at this energy. Each solution Ψ_q is characterized by a vector of coefficients α_{ipq} ($i = 0, 1$) defined as in Eq. (1.5) by the asymptotic forms of the open-channel radial wave functions. A rectangular matrix α can be defined as the $2n_C \times n_C$ array of elements α_{ipq} forming n_C linearly independent column vectors, one for each independent degenerate solution. Any nonsingular linear transformation of these vectors produces a physically equivalent set of solutions. Hence a given rectangular matrix α can be multiplied on the right by any $n_C \times n_C$ nonsingular matrix to produce equivalent solutions. The canonical form

$$\alpha_0 = I, \qquad \alpha_1 = K \tag{1.16}$$

defines the *reactance* matrix K. Here a condensed matrix notation is used with open-channel indices p, q suppressed and with matrices and vectors segmented according to the indices $i = 0, 1$ defined by Eq. (1.5). Matrix α_0 is the $n_C \times n_C$ square matrix upper half of α, and α_1 is the lower half, while I is the $n_C \times n_C$ unit matrix. An arbitrary solution matrix α can be reduced to the form of Eq. (1.16) by multiplying on the right by α_0^{-1} if α_0 is not singular. In general,

$$K = \alpha_1 \alpha_0^{-1}. \tag{1.17}$$

Alternatively, if α_1 is not singular,

$$K^{-1} = \alpha_0 \alpha_1^{-1}. \tag{1.18}$$

For exact solutions, K is real and symmetric (Mott and Massey, 1965; Newton, 1966). It can be diagonalized by an orthogonal transformation. The transformation coefficients define *eigenchannel* vectors, and the corresponding eigenvalues define the quantities

$$\tan \eta_\sigma, \qquad \sigma = 1, \ldots, n_C, \tag{1.19}$$

where η_σ is an *eigenphase*. This representation is useful in computing matrix functions of K by applying the inverse orthogonal transformation defined by the eigenchannel vectors to the diagonal matrix defined by the required function of the eigenphases.

The *scattering* matrix S is defined (Mott and Massey, 1965) by

$$S = (I + iK)(I - iK)^{-1}, \tag{1.20}$$

and the *transition* matrix T can be defined by

$$T = \frac{1}{2i}(S - I) = K(I - iK)^{-1}. \tag{1.21}$$

In terms of eigenchannels and eigenphases,

$$K_{pq} = \sum_{\sigma=1}^{n_C} x_{p\sigma} x_{q\sigma} \tan \eta_\sigma, \qquad\qquad p, q = 1, n_C, \qquad (1.22)$$

$$S_{pq} = \sum_{\sigma=1}^{n_C} x_{p\sigma} x_{q\sigma} \exp(2i\eta_\sigma), \qquad\qquad p, q = 1, n_C, \qquad (1.23)$$

and

$$T_{pq} = \sum_{\sigma=1}^{n_C} x_{p\sigma} x_{q\sigma} \exp(i\eta_\sigma) \sin \eta_\sigma, \qquad p, q = 1, n_C. \qquad (1.24)$$

The S matrix is obviously unitary and symmetric, while the T matrix is symmetric. This particular definition of the T matrix reduces most directly for scattering by a central potential to the phase shift factor appearing in the scattering amplitude,

$$f(\theta) = \frac{1}{k} \sum_{l=0}^{\infty} (2l + 1) \exp(i\eta_l) \sin \eta_l P_l(\cos \theta), \qquad (1.25)$$

defined such that the differential cross section is

$$d\sigma/d\Omega = |f(\theta)|^2. \qquad (1.26)$$

If k is in atomic units a_0^{-1}, the differential cross section is given in units a_0^2 per steradian.

General formulas for differential and total cross sections have been derived for multichannel scattering by Blatt and Biedenharn (1952) and by Jacob and Wick (1959). The partial cross section for scattering from channel q to channel p is

$$\sigma_{qp} = \frac{4\pi}{k_q^2} |T_{pq}|^2 \qquad (1.27)$$

$$= \frac{4\pi}{k_q^2} \left| \sum_\sigma x_{p\sigma} x_{q\sigma} \exp(i\eta_\sigma) \sin \eta_\sigma \right|^2. \qquad (1.28)$$

The total cross section for unpolarized scattering is obtained by summing σ_{qp} over degenerate final states and by averaging over initial states. For scattering by a central potential in partial wave l, this introduces a degeneracy factor $2l + 1$ in Eq. (1.28). Equivalently, the cross section can be summed over all intermediate states of the total scattering system and divided by the degeneracy of the initial state. In LS coupling, the T matrix is diagonal in the total quantum numbers $LS\pi$ and independent of total M_L and M_S. Hence the sum is over $LS\pi$ only, with each term weighted by the degeneracy factor $(2L + 1)(2S + 1)$. The degeneracy of an initial state $(LS\pi)_\gamma$ of the target atom, including a factor of 2 for the spin degeneracy of the incident electron, gives a factor $2(2L_\gamma + 1)(2S_\gamma + 1)$ in the denominator. Hence the total cross section for transition $\gamma \to \gamma'$ is

$$\sigma_{\gamma\gamma'} = \frac{4\pi}{k_\gamma^2} \sum_{L,S,\pi} \frac{(2L + 1)(2S + 1)}{2(2L_\gamma + 1)(2S_\gamma + 1)} \sum_{l,l'} |T_{\gamma'l',\gamma l}^{LS\pi}|^2, \quad (1.29)$$

where l, l' are orbital angular momentum quantum numbers of the incident and scattered electron, respectively.

The differential cross section for scattering from state $\gamma\mu$ (where μ denotes M_L, M_S, m_s) to state $\gamma'\mu'$, for electron deflection angles θ, ϕ, is

$$\frac{d\sigma_{\gamma\mu,\gamma'\mu'}}{d\Omega} = |(\gamma'\mu'|f(\theta, \phi)|\gamma\mu)|^2, \quad (1.30)$$

where (Blatt and Biedenharn, 1952)

$$(\gamma'\mu'|f|\gamma\mu) = \frac{1}{k_\gamma} \sum_{LS\pi} \sum_{ll'} [(2l' + 1)(2l + 1)]^{1/2} i^{l-l'} d_{m'0}^{l'}(\theta) \exp(im'\phi) T_{\gamma'l',\gamma l}^{LS\pi}$$

$$\times (L_{\gamma'}M_{L\gamma'}l'm'|LM_L)(L_\gamma M_{L\gamma}l0|LM_L)$$

$$\times (S_{\gamma'}M_{S\gamma'}\tfrac{1}{2}m_s'|SM_S)(S_\gamma M_{S\gamma}\tfrac{1}{2}m_s|SM_S). \quad (1.31)$$

Here the angular factor is

$$d_{m'0}^{l'}(\theta) \exp(im'\phi) = \left(\frac{4\pi}{2l' + 1}\right)^{1/2} Y_{l'm'}(\theta, \phi) \quad (1.32)$$

and

$$M_L = M_{L\gamma} = M_{L\gamma'} + m', \qquad M_S = M_{S\gamma} + m_s = M_{S\gamma'} + m_s'. \quad (1.33)$$

The total differential cross section for the process $\gamma \to \gamma'$ is

$$\frac{d\sigma_{\gamma\gamma'}}{d\Omega} = \frac{1}{2(2L_\gamma + 1)(2S_\gamma + 1)} \sum_{\mu'} \sum_{\mu} |(\gamma'\mu'|f|\gamma\mu)|^2. \quad (1.34)$$

Explicit formulas for the sums appearing in Eq. (1.34), expressed in compact form by use of angular momentum recoupling coefficients, are given by Blatt and Biedenharn (1952). When contributions of a large number of partial waves must be combined, it may be more convenient to compute the complex amplitudes of Eq. (1.31) directly.

1.3. Close-Coupling Expansion

The basic idea of the close-coupling formalism (Seaton, 1953; Percival and Seaton, 1957; Burke, 1965, 1968; Burke and Seaton, 1971; Smith,

1971; Eissner and Seaton, 1972; Seaton, 1973) is that the set of target states Θ_p included in Eq. (1.1) can be extended to completeness. In this limit, the final term would be unnecessary. The radial factors of the channel orbitals ψ_{ps} are determined by a set of coupled integrodifferential equations, referred to as the *close-coupling* equations. The principal difficulty is that a complete set of target atom eigenstates must include the ionization continuum. The channel index p becomes a continuous variable, and the number of coupled equations is not denumerable.

In practice, only a relatively small number of coupled equations can be solved, so the target states Θ_p included in Eq. (1.1) must be carefully selected. Virtual target excitation into the ionization continuum must be approximated by inclusion of closed-channel pseudostates that cannot be target eigenstates but must be thought of as wave packets in the continuum. Target atom polarization response to the incident electron is treated by the use of polarization pseudostates $\Theta_{\gamma(p)}$, to be discussed in Section 1.4.

For atoms other than hydrogen or one-electron ions, the target atom states are not known exactly. Variational approximations must be used for these states, which in general are expressed as linear combinations of N-electron Slater determinants or LS eigenfunctions constructed from a specified list of orbital functions. It is found to be convenient to require the channel orbital functions ψ_{ps} to be orthogonal to all of these basis orbitals. To compensate for this constraint, $(N + 1)$-electron functions Φ_μ constructed from the basis orbitals must be included in the form of the final term in Eq. (1.1) (Norcross, 1969; Eissner and Seaton, 1972; Seaton, 1973). Thus, the general form of Eq. (1.1) with the orthogonality conditions described above is basic to both the close-coupling expansion and the direct variational methods to be described here.

Since the functions $\{\Phi_\mu\}$ define a Hilbert space, the intuitively obvious equation to use for determining the coefficients $c_{\mu s}$ is the projection of the Schrödinger equation into that space

$$(\Phi_\mu | H - E | \Psi_s) = 0, \qquad \text{all } \mu, \tag{1.35}$$

or

$$\sum_\nu (H - E)_{\mu\nu} c_{\nu s} = -\sum_p (\Phi_\mu | H - E | \mathscr{A} \Theta_p \psi_{ps}). \tag{1.36}$$

If there were no orthogonality constraints, the corresponding equation for determining the channel orbital functions would be

$$(\Theta_p | H - E | \Psi_s) = 0, \qquad \text{all } p. \tag{1.37}$$

Since Θ_p is an N-electron function while Ψ_s is an $(N + 1)$-electron function, this gives a set of coupled integrodifferential equations. This system is reduced to the usual close-coupling equations by projection onto orbital angular momentum states and by including Lagrange multipliers required by the orthogonality conditions. The systems of Eqs. (1.36) and (1.37) are solved simultaneously (Seaton, 1974).

Equation (1.36) determines the coefficients $c_{\mu s}$ as linear functionals of the channel orbitals ψ_{ps}. If these functionals are substituted back into Eq. (1.37), this is equivalent to using the operator M'_{PP} as in Eq. (1.12) or using the one-electron operator m^{pq} defined by Eq. (1.15) in the close-coupling equations. Hence the physical effect of the final term in Eq. (1.1) is to introduce a general nonlocal optical potential into modified close-coupling equations. This potential represents energy-dependent effects of mutual polarization and correlation of the external electron and the target atom.

In terms of the operator m^{pq}, the coupled equations for the channel orbitals follow from Eq. (1.37). This gives the generalized close-coupling equations for radial functions $f_{ps}(r)$ constrained to be orthogonal to the set of basis functions $\{\eta_{pa}(r)\}$ for orbitals with angular momentum l_p,

$$\sum_q m^{pq} f_{qs} = \sum_a \eta_{pa} \lambda^p_{as}. \tag{1.38}$$

The index s labels independent degenerate solutions of these equations, one for each open scattering channel. From Eq. (1.13), the optical potential term in m^{pq} is obtained by integrating

$$\sum_\mu (\Theta_p | H - E | \Phi_\mu) c_{\mu s} \tag{1.39}$$

over orbital angular momentum and spin factors to produce a function of the form

$$-\sum_\mu U_{p\mu}(r) c_{\mu s}. \tag{1.40}$$

Then Eq. (1.38) can be written as

$$\sum_q m^{pq}_{(0)} f_{qs} = \sum_a \eta_{pa} \lambda^p_{as} + \sum_\mu U_{p\mu} c_{\mu s}, \tag{1.41}$$

where $m^{pq}_{(0)}$ excludes the matrix optical potential. Any solution of Eq. (1.41) can be expressed in the form

$$f_{qs} = f^{(0)}_{qs} + \sum_{p,a} f^{(a)}_{qp} \lambda^p_{as} + \sum_\mu f^{(\mu)}_q c_{\mu s}, \tag{1.42}$$

where

$$\sum_q m^{pq}_{(0)} f^{(0)}_{qs} = 0, \qquad \text{all } p, s,$$

$$\sum_q m^{pq}_{(0)} f^{(a)}_{qs} = \eta_{pa} \delta_{ps}, \qquad \text{all } a, p, s, \tag{1.43}$$

$$\sum_q m^{pq}_{(0)} f^{(\mu)}_q = U_{p\mu}, \qquad \text{all } \mu, p.$$

Since the coefficients $c_{\mu s}$ and the Lagrange multipliers λ_{as}^{p} occur linearly in f_{qs} or ψ_{qs}, the equations

$$\sum_{\nu} (\Phi_{\mu}|H - E|\Phi_{\nu})c_{\nu s} = -(\Phi_{\mu}|H - E|\sum_{q} \mathscr{A}\Theta_{q}\psi_{qs}), \qquad \text{all } \mu,$$

$$(\eta_{qa}|f_{qs}) = 0, \qquad \text{all } a, q, \qquad (1.44)$$

are a set of linear inhomogeneous equations that determine these coefficients and Lagrange multipliers. Their values can be computed after Eqs. (1.43) are integrated and substituted into Eq. (1.42) to give a particular solution of Eq. (1.38). Boundary conditions on $f_{qs}^{(0)}$ for large r are chosen to define independent solutions.

These equations show that each $(N + 1)$-electron basis function Φ_{μ} included in Ψ_Q adds one or more inhomogeneous equations to the set that must be integrated but does not increase the dimension of the homogeneous equation system for $f_{qs}^{(0)}$. The system of Eqs. (1.43) can be solved non-iteratively (Burke and Seaton, 1971; Seaton, 1973). Despite these simplifications, computations become unwieldy as the basis set $\{\Phi_{\mu}\}$ is increased or as more open channels are included, and more powerful methods would be desirable.

1.4. Polarization Potentials and Pseudostates

For a neutral atom in a spherically symmetric state (S state), the electric dipole polarization potential is the term of longest range in the scattering potential. This term dominates low-energy electron scattering. The polarization potential, an effective local potential defined in the limit of large radial coordinate r by the optical potential in the operator m^{pq} defined by Eq. (1.15), is due to mutual electric dipole polarization of the target atom and the external electron.

A dipole polarization pseudostate $\Theta_{\gamma(p)}$ is defined as the normalized first-order perturbing wave function computed for target state Θ_p in a perturbing external uniform electric field (Damburg and Karule, 1967; Damburg and Geltman, 1968; Burke *et al.*, 1969b; Geltman and Burke, 1970; Vo Ky Lan *et al.*, 1976). The effective potential acting at large r in the modified close-coupling equations, due to inclusion of $\Theta_{\gamma(p)}$ as a closed-channel pseudostate, can be derived by a spherical harmonic analysis of the resulting optical potential, if $\mathscr{A}\Theta_{\gamma(p)}\psi_{\gamma(p)}$ is treated as a component Φ_{μ} of Ψ_Q in Eqs. (1.1), (1.9), and (1.13).

When expanded in the spherical polar coordinates of two electrons, the Coulomb interaction potential $1/r_{12}$ depends on the radial coordinates through a factor

$$r_<^\lambda / r_>^{\lambda+1} \tag{1.45}$$

multiplying spherical harmonics of degree λ in the angular variables. Here $r_<$ is the lesser of r_1, r_2 and $r_>$ is the greater. For a neutral target atom, terms with $\lambda = 0$ from the electronic Coulomb interaction cancel against the Coulomb potential of the nucleus, so there is no long-range monopole potential. If the atom is in a nonspherical state $(L > 0)$, diagonal matrix elements of the Coulomb interaction occur for even λ such that $0 < \lambda \le 2L$. From Eq. (1.45), this produces external multipole potential terms proportional to $1/r^{\lambda+1}$. When L is not zero, the dominant term is the electric quadrupole potential, which is proportional to r^{-3}.

Off-diagonal matrix elements of spherical harmonic λ connect a target atom channel state of given L to pseudostates with

$$L' = |L - \lambda|, |L - \lambda| + 1, \ldots, L + \lambda. \tag{1.46}$$

If such pseudostates are included as explicit closed channels in the close-coupling equations, the off-diagonal potential connecting the given channel state with these pseudostates is asymptotically proportional to $1/r^{\lambda+1}$. If the partitioning transformation of Eq. (1.12) is used to replace the closed-channel pseudostates by the equivalent optical potential, this off-diagonal matrix element contributes quadratically to an effective polarization potential acting on the given channel state. Hence the effective potential is asymptotically proportional to $1/r^{2\lambda+2}$. When $\lambda = 1$, detailed analysis shows that in the limit of large r this effective potential is the polarization potential $-\alpha/2r^4$, where α is the static electric dipole polarizability of the target atom in the given channel state. The argument outlined here can be made more precise by considering the explicit asymptotic expansion of the radial factor of the pseudostate channel orbital $\psi_{\gamma(p)}$ as determined by the close-coupling equations (Burke and Schey, 1962). In general, virtual excitations of the target atom with multipole index λ produce a multipole polarization potential of the asymptotic form $-\alpha_\lambda/2r^{2\lambda+2}$, where α_λ is a multipole polarizability.

The use of polarization pseudostates in the close-coupling formalism greatly simplifies the treatment of the polarization potential but still involves essential approximations. In particular, the pseudostate function $\Theta_{\gamma(p)}$ is computed for a static perturbing potential in the form of a uniform electric field, valid in principle only when the scattered electron is completely outside the target state charge distribution and when the perturbation is constant in time. Penetration of the external electron inside the target charge distribution should change the functional form of $\Theta_{\gamma(p)}$, not just multiply it by a numerical coefficient as is implicit in the antisymmetrized product form $\mathscr{A}\Theta_{\gamma(p)}\psi_{\gamma(p)}$ used in the close-coupling expansion. A penetration effect that alters the functional form of $\Theta_{\gamma(p)}$ for different values of the radial coordinate of the external electron requires a summation such as

$$\sum_a \mathscr{A}\Theta_{\gamma(p)a}\psi_{\gamma(p)a}. \qquad (1.47)$$

Full treatment of this effect within the close-coupling formalism would require the inclusion of a sequence of pseudostate components $\Theta_{\gamma(p)a}$ as independent closed-channel functions. Each added channel orbital $\psi_{\gamma(p)a}$ increases the number of close-coupling equations to be solved.

The polarization response of an atom to an external potential is described in general by the frequency-dependent polarizability $\alpha(\omega)$. If only one pseudostate $\Theta_{\gamma(p)}$ is included in the close-coupling expansion, it can correspond to only a particular value of the frequency ω. In all existing applications to electron–atom scattering, the static polarizability $\alpha(0)$ has been used. Following the derivation given above, analogy between the energy dependence of the optical potential and the theory of ordinary optical excitation indicates that the correct choice of effective frequency should be

$$\hbar\omega = \Delta E, \qquad (1.48)$$

where ΔE is the energy transferred during the scattering process. For elastic scattering no energy is transferred, so the use of pseudostates appropriate to $\alpha(0)$ is justified; but different functions may be required for inelastic scattering. In addition, in the case of inelastic scattering, polarization pseudostates must be included independently for each distinct open-channel target state, further adding to the complexity of the close-coupling equations.

In considering the treatment of the polarization potential, the close-coupling formalism with polarization pseudostates can be compared, for elastic scattering, with the polarized orbital method (Temkin, 1957; Temkin and Lamkin, 1961; Drachman and Temkin, 1972), in which an effective local potential is explicitly introduced as a model of the true optical potential acting in a single open channel. Singlet and triplet s-wave phase shifts for elastic e^-–H scattering, computed in various versions of the polarized orbital method and in the pseudostate close-coupling formalism, have been compared by Callaway (1973) with accurate phase shifts computed variationally (Schwartz, 1961b; Armstead, 1968; Gailitis, 1965b).

In general, this comparison shows that the polarized orbital single-channel approximation is somewhat less successful than the pseudostate close-coupling method, which gives excellent results. It can be concluded that errors due to the penetration effect are not large, at least for low scattering energies. Since the polarized orbital method is not formulated for multichannel scattering, no similar comparison is possible in that case.

1.5. Continuum Bethe–Goldstone Equations

The philosophy behind the use of the close-coupling expansion is to make the first term of Eq. (1.1),

$$\Psi_P = \sum_p \mathcal{A}\Theta_p\psi_{ps}, \tag{1.49}$$

as complete as possible. The use of polarization pseudostates as closed-channel functions makes it possible to represent the physical effect of a polarization potential while keeping the number of coupled equations small. However, because the number of equations grows with each additional open or closed channel included in Ψ_P, there is a severe practical limitation on the number of these channels and hence on the convergence of the close-coupling expansion as a representation of the true scattering wave function Ψ_s. In order to describe inelastic collision processes coupling several target atom states and to allow for scattering by atoms with open-shell ground states, a more efficient method of treating polarization and correlation effects is desirable. An efficient method is also needed for quantitative studies of convergence since the pseudostate representation is itself only an approximation to a more general expression in independent components of the closed-channel functions.

An alternative approach is to recognize that the second term of Eq. (1.1),

$$\Psi_Q = \sum_\mu \Phi_\mu c_{\mu s}, \tag{1.50}$$

can be made as complete as possible within the limits of quadratic integrability. In this context, the usual idea of completeness as applied to bound-state wave functions is inadequate since no basis of quadratically integrable functions can represent the open-channel terms in Ψ_s. However, if the open-channel orbitals ψ_{ps} are specified in the asymptotic limit $r \to \infty$, they can be represented by functions that are regular at the origin and approach the given asymptotic form but are otherwise arbitrary. This arbitrariness can be compensated for by inclusion of appropriate quadratically integrable terms in Ψ_Q. Thus, with regard to Ψ_Q, the essential requirement of completeness is to represent the localized portion of the scattering wave function, Ψ_s, excluding its asymptotic open-channel component. Since this can be done with quadratically integrable functions, the expansion of Ψ_Q in basis functions is a problem of the same kind as that encountered in bound-state calculations, where computational methods of considerable sophistication have been developed for its practical solution. By shifting the burden of completeness from Ψ_P to Ψ_Q, these methods become available for use in the study of electron scattering by atoms and molecules.

The exact function to be represented by Ψ_Q is not uniquely defined unless the open-channel orbitals ψ_{ps} are specified. Even given the exact scattering wave function Ψ_s, one could continue indefinitely to add basis functions to Ψ_Q and to orthogonalize the functions $\mathscr{A}\Theta_p\psi_{ps}$ to them. This indicates that a meaningful convergence criterion cannot be provided by considering the function Ψ_Q by itself. Instead, one must examine the asymptotic limit of the open-channel orbitals, which is determined by the coefficients α_{0ps} and α_{1ps} in Eq. (1.5). The stability of these coefficients, or of the scattering matrices defined by them, against addition of basis functions to the Hilbert space $\{\Phi_\mu\}$ is the ultimate test of convergence. This structural feature of the scattering wave function lies behind the variational methods valid in this theory, all of which are based on a stationary property of some expression for the coefficients α_{ips}.

A systematic method is needed for constructing and extending the Hilbert space basis for Ψ_Q, and a procedure must be specified to compute the asymptotic coefficients α_{ips} for any given basis $\{\Phi_\mu\}$. In principle, from a given orbital basis $\{\phi_i; \phi_a\}$ one could construct all possible $(N + 1)$-electron Slater determinants, as defined in Eq. (1.7), or the corresponding LS eigenfunctions. In bound-state calculations, this would be called a full CI (configuration interaction) calculation in the given orbital basis. In practice, the number of Slater determinants increases very rapidly with the number of target atom electrons and with the number of basis orbital functions. Except for the lightest atoms, this implies that calculations requiring a full configuration interaction expansion cannot be carried to convergence in terms of basis orbitals unless some simplifications are introduced.

For bound-state calculations, a particular approach to this problem makes use of a hierarchy of Bethe–Goldstone equations in matrix variational form (Nesbet, 1969b). This method has been used successfully for calculations of bound-state energies and of hyperfine structure parameters. The generalization appropriate to the scattering problem (Nesbet, 1967) will be described here.

The Bethe–Goldstone equation, which originated in nuclear many-body theory (Brueckner, 1959; Bethe and Goldstone, 1957), is the Schrödinger equation for a pair of particles embedded in the Fermi sea of the remaining $N - 2$ particles in an N-particle system. The two-particle Bethe–Goldstone wave function is constrained to be orthogonal to $N - 2$ specified occupied orbitals of the Fermi sea. This is an independent pair model in which the correlation energy of each pair of particles is computed exactly. The total correlation energy is estimated by adding the pair correlation energies (Gomes *et al.*, 1958). Mittleman (1966) proposed that the same approach could be used in electron–atom scattering theory. In particular, scattering of an electron by an alkali atom can be approximated by considering a continuum Bethe–Goldstone electron-pair wave function for the external electron and the valence electron, constrained only by ortho-gonality to the occupied orbitals of the closed-shell positive ion core. This wave function satisfies boundary conditions at large r appropriate to one bound electron and one free electron of specified asymptotic momentum. Mittleman proposed that the continuum Bethe–Goldstone equation should be solved directly as a partial differential equation using a method developed by Temkin (1962).

For an N-electron alkali atom, the component Ψ_Q of the $(N + 1)$-electron scattering wave function considered by Mittleman can be expanded as a linear combination of Slater determinants $\{\Phi^a, \Phi_i^{ab}\}$ in the notation of Eq. (1.7). Here index i has a specific value, referring to the *ns* valence electron of the target atom. This set of Slater determinants defines a *variational subspace* of the $(N + 1)$-electron Hilbert space $\{\Psi_Q\}$. This subspace can be denoted by $[i]$ for the occupied orbital whose virtual excitations define it. As the orbital set $\{\phi_i; \phi_a\}$ becomes complete, a variational scattering calculation using the Hilbert space $[i]$ becomes equivalent to solution of the continuum two-electron Bethe–Goldstone equation.

The physical significance of individual Slater determinants in $[i]$ can be understood in terms of the discussion given above. For elastic scattering, if the target atom ground state is approximated by a single Hartree–Fock Slater determinant Φ_0, the functions Φ^a as defined by Eq. (1.7) contribute only to representation of the inner part of the open-channel orbital ψ_{ps}. Each Φ^a is constructed by antisymmetrizing a quadratically integrable orbital ϕ_a into the unmodified target atom function Φ_0. A calculation with Hilbert space component Ψ_Q restricted to expansion in the basis $\{\Phi^a\}$, to be denoted by $[0]$ here, corresponds to the *static exchange* approximation. There are no virtual excitations of the target atom, and there is no polarization potential.

A Slater determinant Φ_i^{ab} as defined by Eq. (1.7) represents a virtual excitation $\phi_i \to \phi_a$, denoted by (a/i), of the target atom, coupled to an additional quadratically integrable orbital ϕ_b. This is exactly the structure of the closed-channel terms $\mathcal{A}\Theta_\gamma\psi_\gamma$ in the close-coupling expansion. In the Hartree–Fock approximation for the target atom state, multipole polarizabilities are described by first-order perturbation wave functions expressed as linear combinations of Slater determinants Φ_i^a, corresponding to one-electron virtual excitations (a/i). A closed-channel pseudostate Θ_γ

represents such a perturbation function. With this interpretation, the additional orbital ϕ_b in Φ_i^{ab} represents a basis function for the linear expansion of an external closed-channel orbital ψ_γ.

In the formalism of the polarized orbital method (Drachman and Temkin, 1972), closed-channel terms in Ψ cannot be expressed in the separable form $\mathscr{A}\Theta_\gamma\psi_\gamma$ assumed in the close-coupling expansion. Inclusion of functions Φ_i^{ab} in Ψ_Q takes this failure of separability into account since the coefficients c_i^{ab} are independent parameters determined by Eq. (1.36). This means that the general nonseparable form, Eq. (1.47), is incorporated into the Bethe–Goldstone approximation.

If the target atom reference state belongs to an open-shell configuration, virtual excitations must be considered in terms of configurations rather than of individual orbitals. In the usual Hartree–Fock approximation for such atomic states (Hartree, 1957), occupied orbitals are grouped into subshells having common radial factors $f_{nl}(r)$. For each radial function, the angular and spin factors give $2(2l + 1)$ orbital functions with different quantum numbers m_l and m_s. Any N-electron Slater determinant constructed from these orbital functions defines a set of subshell occupation numbers

$$d_{nl} \le 2(2l + 1) \qquad (1.51)$$

that add up to N, specifying the number of occupied orbitals with the same $f_{nl}(r)$ factor. Slater determinants with the same occupation numbers are said to belong to the same configuration. As used here, *configuration* refers to such a set of Slater determinants or to the linear function space spanned by them as basis functions. Since the angular momentum operators $\mathbf{L}^2$ and $\mathbf{S}^2$ can be diagonalized without mixing different configurations, LS eigenfunctions can also be characterized as belonging to a particular configuration.

For open-shell atoms, it is convenient to use LS eigenfunctions for the construction of Ψ_Q and to define the structure of the variational Hilbert space in terms of configurational virtual excitations. Then the notation a/i for a single virtual excitation with respect to a reference configuration means that the occupation number d_a is increased by one and the occupation number d_i is decreased by one with reference to subshells $(nl)_a$ and $(nl)_i$, respectively. It is assumed that quantum numbers L and S are specified and that all possible linearly independent LS eigenfunctions that can be constructed from a given configuration are included in the variational basis set if that configuration is included.

When the Bethe–Goldstone approximation is characterized in terms of configurational virtual excitations, the variational space denoted by $[i]$ replaces the reference determinant Φ_0 by a set of Slater determinants spanning the reference state configuration, and each virtually excited determinant such as Φ^a or Φ_i^{ab} is accompanied by all other determinants of the same configuration compatible with specified total angular momentum quantum numbers L, S, M_L, and M_S.

From the analysis given here, it is clear that for a single configuration target state, the physical effects of a generalized multipole polarization potential can be described by a wave function constructed from determinants of the form Φ^a and Φ_i^{ab} or from the corresponding configurations. Assuming a Hartree–Fock target state (single configuration), the functions $\{\Phi^a\}$ give the static-exchange expansion of an open-channel orbital for elastic scattering, and the functions $\{\Phi_i^{ab}\}$ describe the coupling of a multipole polarization or virtual excitation of the target atom to a virtual excitation of the external electron.

The direct solution of the Bethe–Goldstone equation as a partial differential equation has been found to be quite difficult compared with the variational methods to be presented in detail below. The replacement of closed-channel terms by an $(N + 1)$-electron configuration interaction expansion, as in Eq. (1.50), shifts the emphasis of computational methods from analytic techniques, the integration of integrodifferential equations, to algebraic techniques, manipulation of matrices and coefficient vectors. Variational methods are particularly well adapted to the algebraic or matrix formulation of the computational problem.

While less promising for general applications than algebraic variational methods, analytic techniques can still be used when closed-channel terms are included entirely in Ψ_Q rather than in Ψ_P. In fact, as mentioned in Section 1.3, some terms of this kind are necessarily included in standard close-coupling calculations since they are required by the orthogonality conditions. Noniterative methods developed for treating these terms (Burke and Seaton, 1971) can easily be applied to the matrix optical potential defined by Eqs. (1.13) and (1.15). This formalism, treating all closed-channel terms through the matrix optical potential, has been proposed as a practical computational methods (Chen and Chung, 1970; Chen et al., 1971) and applied to calculations of e^-–H scattering resonances.

1.6. Generalizations of the Bethe–Goldstone Approximation

When the target state variational space is represented by more than a single configuration, either to describe target states for several open scattering channels or to include target atom correlation effects, the variational basis for the scattering wave function must be correspondingly augmented. In the close-coupling expansion, consistency of the orthogonality conditions requires that for each N-electron wave function Θ_p included in Ψ_P of Eq. (1.49), all possible $(N + 1)$-electron functions of the form

$$\mathscr{A}\Theta_p\phi_k \tag{1.52}$$

compatible with total quantum numbers $LS\pi$ must be included in Ψ_Q. Here ϕ_k is any orbital in the basis set $\{\phi_i; \phi_a\}$. In the alternative methods considered here, the variational basis for the states Θ_p is a specified list of N-electron configurations. The required consistency condition is met if the basis for Ψ_Q includes all $(N + 1)$-electron configurations formed from the $\{\Theta_p\}$ list by adding one orbital. This is sufficiently general to apply to the Bethe–Goldstone approximation and to various generalizations considered here. When this condition is imposed, both N-electron and $(N + 1)$-electron variational bases are determined by specifying only this list of configurations. This structure will be assumed here unless otherwise stated.

The concept of the Bethe–Goldstone equation can be generalized to provide a step-by-step procedure which, if it could be implemented, would converge in a finite number of steps to exact results for scattering by any N-electron atom (Nesbet, 1967). Each computational step involves the solution of generalized continuum Bethe–Goldstone equations of order n, defined to be equivalent to a variational calculation in which all possible configurational virtual excitations of n orbitals $\phi_i, \phi_j, \ldots$, specified by their radial subshell quantum numbers, are represented in the variational wave function. The corresponding Hilbert space of Ψ_Q will be denoted by $[ij\ldots]$, indicating the n indices $i, j, \ldots$ for orbitals or subshells occupied in a reference configuration. This Hilbert space is a direct sum of *disjoint* subspaces $(ij\ldots)$, where this notation indicates the linear space spanned by all configurations obtained from the reference configuration (0) by virtual excitations from occupied subshells with the specified indices to subshells unoccupied or not completely occupied in (0). A *variational* Hilbert space $[ij\ldots]$ is defined as the direct sum of all disjoint subspaces whose indices form a subset of $ij\ldots$. For example,

$$[ij] = (0) + (i) + (j) + (ij), \tag{1.53}$$

where the notation signifies a direct sum of vector spaces.

The complete Hilbert space for $(N + 1)$-electron wave functions is the direct sum of all disjoint subspaces with up to N indices,

$$[\Psi_Q] = (0) + \sum_i (i) + \sum_{ij} (ij) + \sum_{ijk} (ijk) + \cdots. \tag{1.54}$$

The variational subspaces form a *lattice* in the technical sense that

$$[0] \subseteq [ijk\ldots] \subseteq [\Psi_Q], \tag{1.55}$$

for any $[ijk\ldots]$. Simple ordering holds for variational subspaces with nested indices,

$$[0] \subset [i] \subset [ij] \subset [ijk\ldots], \tag{1.56}$$

but no ordering relation is defined otherwise. For configurational virtual excitations, the orbital indices here may be repeated since they refer to the occupied subshells of the reference configuration. The indexing is unique if, in $[ijk\cdots]$,

$$i \le j \le k \le \cdots, \tag{1.57}$$

and the individual indices denote l, n treated as a two-digit number.

This lattice structure of Hilbert spaces defines a hierarchy of variational calculations to be carried out independently for each different variational subspace corresponding to a node in the lattice diagram. The useful information gained by each such calculation is defined inductively in terms of *net increments* of properties of the variational wave function. Values of such properties computed directly for a given variational space $[ij\ldots]$ are defined as *gross increments* $\Delta F_{ij\ldots}$, usually obtained as corrections to properties of the reference state or of a static-exchange calculation in scattering theory. A *net increment* $f_{ij\ldots}$ is defined as the difference between the corresponding gross increment and the sum of all net increments of lower order whose indices form a proper subset of the given set $ij\ldots$. This definition of gross and net increments in terms of a lattice decomposition of the variational Hilbert space ensures that when each gross increment calculation is carried to convergence a given physical property is expressed exactly by the finite sum of all net increments up to order N. Summations of net increments through lower-order n define a converging sequence of approximations. In bound-state calculations, correlation energies are often obtained to satisfactory accuracy by terminating this summation at order 2, corresponding to the independent pair approximation (Nesbet, 1969b). It can be shown that each net increment is equivalent to a sum to infinite order of a certain subset of perturbation theory diagrams (Nesbet, 1970).

As a concrete example of this structure and notation, consider electron scattering by atomic oxygen. The ground-state configuration is $1s^2 2s^2 2p^4$ with three distinct states ^3P, ^1D, and ^1S. At energies above $E(^1S)$ but below the next excitation threshold, each of these states defines open scattering channels. A reactance matrix $K^{LS\pi}$ is to be computed for each possible set of total quantum numbers. With respect to the ground-state configuration as reference, denoted equivalently by (0) or $[0]$, configurational virtual excitations define a lattice of variational subspaces as follows, omitting virtual excitations of the $1s^2$ subshell:

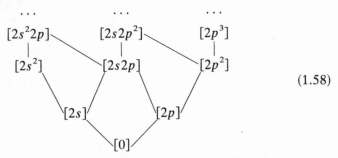

$$(1.58)$$

If gross or net increments of the K matrix are denoted by $K[ij\ldots]$ or $K(ij\ldots)$, respectively, a calculation at level $[2p]$ would give the net increment

$$K(2p) = K[2p] - K(0),$$

where $K(0)$ or $K[0]$ refers to the K matrix computed in a single-configuration calculation. A calculation at level $[2s^2 2p]$ would give the net increment

$$K(2s^2 2p) = K[2s^2 2p] - K(2s^2) - K(2s2p) - K(2s) - K(2p) - K(0).$$

$$(1.59)$$

At order $n = 2$, including all double virtual excitations, the implied approximation is

$$K \cong K(0) + K(2s) + K(2p) + K(2s^2) + K(2s2p) + K(2p^2).$$

$$(1.60)$$

The Bethe–Goldstone approximation, at order 1, is

$$K \cong K(0) + K(2s) + K(2p),$$ $$(1.61)$$

neglecting $1s$ virtual excitations as before.

If the hierarchy of variational calculations is truncated at the level of single subshell excitations $[i]$, as in Eq. (1.61), this corresponds to solution of two-electron continuum Bethe–Goldstone equations, independently for each subshell, adding the incremental contributions to computed quantities such as the K matrix. Following the discussion in Section 1.5, the essential approximation is that polarizability contributions of different subshells are assumed to be independent and additive. Since, essentially, this approximation is involved in the usual perturbation theory of static polarizabilities, it can be expected to give results of useful accuracy in scattering theory. This approximation is inherent in the polarized orbital method and in the use of uncorrelated polarization pseudostates in close-coupling calculations. No calculations to date with the continuum Bethe–Goldstone formalism have gone beyond this order of approximation. Inclusion of higher-order terms would allow for systematic correction of the assumption of additive polarizabilities.

The structure indicated in Eq. (1.61) is typical of the Bethe–Goldstone approximation for an open-shell target atom expressed in terms of virtual excitations. This structure implies two separate calculations, with variational spaces, as defined by Eq. (1.53),

$$[2s] = (0) + (2s),$$
$$[2p] = (0) + (2p). \qquad (1.62)$$

In terms of orbital virtual excitations, starting from the reference state $(1s^2 2s^2 2p^4)^3 P$, with $M_L = 1$, $M_S = 1$, the variational spaces are

$$[2s] = (0) + (2s\beta) + (2s\alpha),$$
$$[2p] = (0) + (2p_1\beta) + (2p_{-1}\alpha) + (2p_0\alpha) + (2p_1\alpha). \qquad (1.63)$$

Here α, β refer to spin $+\frac{1}{2}\hbar$, $-\frac{1}{2}\hbar$, respectively. There is a certain inconsistency in treating virtual excitations of the four distinct $2p$ orbitals and of the two $2s$ orbitals separately. This includes interaction terms between distinct $2p$ virtual excitations in the Hamiltonian matrix and those between distinct $2s$ virtual excitations but omits coupling between $2s$ and $2p$. A more consistent approximation is indicated by

$$[2s2p]_1 = (0) + (2s) + (2p), \qquad (1.64)$$

combining all the disjoint subspaces indicated in Eq. (1.63). The notation introduced here is

$$[ij \cdots]_n = (0) + (i) + (j) + \cdots, \qquad (1.65)$$

including all subsets of indices up to n in number. This level of approximation $[ij\ldots]_1$, has been used for calculations of electron scattering by open-shell atoms. The variational space $[2s2p]_1$ can be included in the lattice diagram of Eq. (1.58) as a proper subspace of $[2s2p]$, itself including $[2s]$ and $[2p]$ as subspaces.

It has been found in bound-state calculations that the Bethe–Goldstone or independent-pair approximation is less satisfactory than coupled-pair approximations in various forms (Cizek and Paldus, 1971; Meyer, 1974; Ahlrichs *et al.*, 1975; Kutzelnigg, 1977). These approximations include terms of the Hamiltonian matrix that couple distinct electron-pair virtual excitations, but modify diagonal elements to account for implied effects of higher-order virtual excitations that can be estimated by use of perturbation theory. Equation (1.64) defines a coupled-pair model for the scattering system but does not take the implied higher-order corrections into account. Coupled-pair approximations have not yet been used in scattering theory, but they may help to define an optimal approximation for open-shell atoms, based on Eq. (1.64).

Formal methods of quantum field theory can be applied to electron–atom scattering. The diagrammatic perturbation theory can be used to define a perturbation expansion of an effective optical potential. In the formal theory, this optical potential appears as the self-energy operator of the one-particle Green's function. The second-order optical potential was computed by Pu and Chang (1966) and used to obtain s- and p-wave e^-–He elastic-scattering phase shifts. Similar calculations were carried out by Kelly (1967, 1968) for singlet and triplet s-wave e^-–H elastic phase shifts. Kelly included corrections for effects higher than second order and obtained results in good agreement with accurate variational values of Schwartz (1961b). Knowles and McDowell (1973) corrected the work of Pu and Chang for some omitted effects and computed e^-–He d-wave phase shifts. The optical potential computed in these applications is formally equivalent to that defined here by Eqs. (1.12) and (1.15), except that the component Ψ_Q of the scattering wave function is expanded in continuum basis functions in the perturbation theory rather than in a discrete quadratically integrable basis.

Another general approach has been developed from the quantum field theoretical formalism of Schwinger (Csanak *et al.*, 1971b; Thomas *et al.*, 1973). This approach develops a hierarchy of coupled equations for n-particle Green's functions (Schwinger, 1951), which have immediate physical significance as transition amplitudes or functions describing the response of a many-particle system to an external perturbation. Truncation of this hierarchy leads to self-consistent approximations at successive levels of complexity. In this aspect, the coupled Green's function theory is similar to the hierarchy of Bethe–Goldstone equations described above.

The formal theory suggests that an approximation referred to as the GRPA (generalized random phase approximation) should contain the essential physics of the polarization response of a target atom to an elastically scattered electron (Schneider *et al.*, 1970). In this approximation, static exchange equations are augmented by an optical potential (response function) constructed from RPA (random phase approximation) transition amplitudes. The GRPA has been applied to e^-–He elastic scattering to obtain s- and p-wave phase shifts in excellent agreement with the best polarized orbital and variational calculations (Yarlagadda *et al.*, 1973). For elastic scattering, the GRPA appears to provide an internally consistent formalism at the same level of physical approximation as the polarized orbital method.

A generalized optical potential for inelastic electron–atom scattering has been derived from the coupled Green's function formalism by Csanak *et al.* (1973). The lowest-order approximation for inelastic scattering is referred to as the RPA S-matrix formula (Csanak *et al.*, 1971a). This approximation has been applied to calculations of electron impact excitation of the $n = 2$ levels of He (Thomas *et al.*, 1974a) in the intermediate energy range (above the ionization threshold). This approximation is remarkably simple, requiring only matrix elements of the RPA transition potential between static-exchange continuum functions for the incident and scattered electron.

Csanak and Taylor (1972, 1973) have analyzed the coupled Green's function formalism in terms of widely used models and approximate methods, including various forms of the polarized orbital method. They show that the Green's function formalism provides a rationale for making internally consistent approximations in such methods.

2

Variational Theory

Introduction

Theoretical computational methods relevant to low-energy electron scattering rely on variational principles. Variational theory appropriate to representation of the $(N + 1)$-electron scattering wave function in the form of Eq. (1.1) will be presented in this chapter. Direct computational application of this formalism is referred to here as the *matrix variational method*. Since it has been shown in Chapter 1 how the widely applied close-coupling method can be described by Eq. (1.1), the general variational argument to be given here also applies to that method, with some modifications of detail. In the *algebraic close-coupling method* this variational formalism, rather than numerical integration, is used to obtain solutions of the coupled equations as linear combinations of basis functions.

In bound-state calculations, the Rayleigh–Ritz variational principle provides both an upper bound to an exact energy and a stationary property that determines free parameters in the wave function. In scattering theory, the energy is specified in advance. Variational principles are used to determine the wave function but do not provide variational bounds, except in special cases. Parameters in the wave function are determined so that a variational functional is made stationary, but the sign of the residual error is not usually determined. This lack of a well-defined bounded quantity is a complicating aspect of calculations in scattering theory since there is no simple absolute standard of comparison between different trial wave functions. The present discussion will be restricted to consideration of stationary estimates of the reactance matrix K. Because the K matrix is real and symmetric, it provides the most convenient of the various alternative representations of the asymptotic part of the scattering wave function.

Variational principles for the K matrix and scattering wave function were originally introduced by Hulthén (1944, 1948) and by Kohn (1948). Developments of the multichannel principle of Kohn form the central part of the material presented here. The variational principle of Schwinger (1947), based on the integral equation formulation of scattering theory, will be presented here in a multichannel form that may provide a practical alternative to methods based on the Kohn principle. Hybrid methods that combine numerical integration and algebraic techniques will be discussed since they can simplify calculations of scattering due to long-range potentials.

Quantitative calculations of electron–atom scattering using variational methods began with calculations of e^-–H elastic phase shifts by Schwartz (1961a, b), using the Kohn method. Further progress required understanding and resolution of the problem of spurious singularities inherent in the Kohn formalism. Several viable alternative methods have been developed and used for a growing number of applications. Review articles describing this work with variational methods have been published by Harris and Michels (1971), by Truhlar *et al.* (1974), by Nesbet (1975b, 1977a), and by Callaway (1978b). Specialized discussions of variational principles in scattering theory are given by Demkov (1963) and by Moiseiwitsch (1966).

2.1. Formalism for Multichannel Scattering

The asymptotic form of radial open-channel orbitals $f_{ps}(r)$ is given by Eq. (1.5) for scattering by a neutral atom. Functions of this kind can be represented as linear combinations of two independent continuum basis functions for each open channel. These radial functions are required to be regular at the origin but to have the asymptotic forms

$$F_{0p}(r) \sim k_p^{-1/2} \sin (k_p r - \tfrac{1}{2} l_p \pi),$$
$$F_{1p}(r) \sim k_p^{-1/2} \cos (k_p r - \tfrac{1}{2} l_p \pi). \tag{2.1}$$

A variational approximation compatible with the required asymptotic form of $f_{ps}(r)$ is

$$f_{ps}(r) = F_{0p}\alpha_{0ps} + F_{1p}\alpha_{1ps}, \tag{2.2}$$

where the coefficients α_{ips} are defined by Eq. (1.5). In the matrix variational method (Nesbet, 1975b, 1977a) the quadratically integrable orbital basis $\{\phi_i; \phi_a\}$ is augmented for each open channel by adding two functions ϕ_{ip} whose radial factors are $r^{-1}F_{ip}(r)$. These functions are orthogonalized to the quadratically integrable basis orbitals $\{\phi_i; \phi_a\}$.

For an exact scattering wave function, the coefficients $c_{\mu s}$ in Eq. (1.1) would be determined by Eq. (1.35) or (1.36) as linear functions of the coefficients α_{ips} since the latter occur linearly in the open-channel orbitals ψ_{ps}. By factoring the coefficients

$$c_{\mu s} = \sum_i \sum_p c_\mu^{ip} \alpha_{ips}, \tag{2.3}$$

the variational approximation to the $(N+1)$-electron scattering wave function of Eq. (1.1) implied by Eq. (2.2) is

$$\Psi_s = \sum_i \sum_p \left(\mathscr{A} \Theta_p \phi_{ip} + \sum_\mu \Phi_\mu c_\mu^{ip} \right) \alpha_{ips}. \tag{2.4}$$

Here, as in the close-coupling Eq. (1.38), index s labels independent degenerate solutions of the continuum Schrödinger equation, one for each open channel at given total energy E.

In the matrix variational method, the coefficients c_μ^{ip} are determined separately, for each set of indices i, p, from the matrix equations

$$\left(\Phi_\mu \middle| H - E \middle| \mathscr{A} \Theta_q \phi_{jq} + \sum_\nu \Phi_\nu c_\nu^{jq} \right) = 0, \qquad \text{all } \mu, j, q, \tag{2.5}$$

which imply Eq. (1.35). Index q refers to open channels only, and H is the $(N+1)$-electron Hamiltonian operator. These equations follow from the variational condition

$$\partial \Xi_{st} / \partial c_\mu^{ip*} = 0, \qquad \text{all } \mu, i, p, \tag{2.6}$$

for arbitrary fixed values of the coefficients α_{ips}. The variational functional, for n_C open-scattering channels, is the $n_C \times n_C$ matrix

$$\Xi_{st} = (\Psi_s | H - E | \Psi_t). \tag{2.7}$$

When Eq. (2.5) is satisfied, the variational functional becomes an explicit quadratic function of the coefficients α_{ips}, which can be assumed to be real numbers,

$$\Xi_{st} = \sum_{ip} \sum_{jq} \alpha_{ips} m_{ij}^{pq} \alpha_{jqt}, \tag{2.8}$$

where, as a consequence of Eq. (2.5),

$$m_{ij}^{pq} = M_{ij}^{pq} - \sum_\mu \sum_\nu M_{ip,\mu} (M^{-1})_{\mu\nu} M_{\nu,jq}. \tag{2.9}$$

The matrices combined in Eq. (2.9) are the *bound–bound* matrix (Hermitian)

$$M_{\mu\nu} = (\Phi_\mu | H - E | \Phi_\nu), \tag{2.10}$$

the *bound–free* matrix (rectangular)

$$M_{\mu,ip} = (\Phi_\mu|H - E|\mathscr{A}\Theta_p\phi_{ip}), \tag{2.11}$$

and the *free–free* matrix (non-Hermitian)

$$M_{ij}^{pq} = (\mathscr{A}\Theta_p\phi_{ip}|H - E|\mathscr{A}\Theta_q\phi_{jq}). \tag{2.12}$$

Phase factors can be chosen so that both M_{ij}^{pq} and m_{ij}^{pq} are real but unsymmetric.

If Ψ_t is an exact solution of the Schrödinger equation

$$(H - E)\Psi_t = 0, \tag{2.13}$$

the coefficients α must satisfy the matrix equations

$$\sum_{jq} m_{ij}^{pq}\alpha_{jqt} = 0, \qquad \text{all } i, p. \tag{2.14}$$

For n_C open channels, these equations have n_C linearly independent solutions, each defined by a column vector with $2n_C$ elements. If the rectangular array of coefficients is segmented as described in Section 1.2, and open-channel indices are suppressed, the reactance matrix K is defined by Eq. (1.17) as $\alpha_1\alpha_0^{-1}$, where α_0 is the square matrix $\{\alpha_{0pq}\}$ and α_1 is the matrix $\{\alpha_{1pq}\}$. The scattering and transition matrices and scattering cross sections are computed from the K matrix as described in Section 1.2.

In this condensed matrix notation, suppressing open-channel indices and segmenting matrices and vectors according to the indices i, j, the variational functional, Eq. (2.8), is

$$\Xi = \alpha^\dagger m\alpha \tag{2.15}$$

$$= \alpha_0^\dagger(m_{00}\alpha_0 + m_{01}\alpha_1) + \alpha_1^\dagger(m_{10}\alpha_0 + m_{11}\alpha_1). \tag{2.16}$$

Equation (2.14), for an exact wave function, is

$$m\alpha = \begin{pmatrix} m_{00} & m_{01} \\ m_{10} & m_{11} \end{pmatrix}\begin{pmatrix} \alpha_0 \\ \alpha_1 \end{pmatrix} = 0. \tag{2.17}$$

Here α is the $2n_C \times n_C$ rectangular array of coefficients α_{ipq}. The symbol $(^\dagger)$ denotes an Hermitian adjoint, or transpose of a real matrix.

In general, for approximate wave functions, the homogeneous system of Eq. (2.17) has no nontrivial solutions. Variational methods are used to obtain an approximate solution matrix α. If α were an exact solution of Eq. (2.17), then the functional Ξ would vanish. An infinitesimal variation of α by $\delta\alpha$ induces a variation of Ξ of the form

$$\delta\Xi = \delta\alpha^\dagger m\alpha + (m\alpha)^\dagger\delta\alpha + \alpha^\dagger(m - m^\dagger)\delta\alpha. \tag{2.18}$$

The last term here does not vanish. The non-Hermitian part of m_{ij}^{pq} comes from the free–free matrix, Eq. (2.12).

When channel orbitals ϕ_{ip} are constructed with radial factors $r^{-1}F_{ip}(r)$ normalized according to Eqs. (2.1), the non-Hermitian part of the free–free matrix is given in atomic units by

$$M_{ij}^{pq} - M_{ji}^{qp} = \tfrac{1}{2}\delta_{pq}(\delta_{i0}\delta_{j1} - \delta_{i1}\delta_{j0}). \qquad (2.19)$$

This formula expresses the surface integral resulting from the action of the kinetic energy operator on open-channel orbitals whose asymptotic forms are specified by Eqs. (2.1). Since the other terms in Eq. (2.9) are Hermitian (symmetric and real by appropriate choice of phase factors), this implies that

$$m_{ij}^{pq} - m_{ji}^{qp} = \tfrac{1}{2}\delta_{pq}(\delta_{i0}\delta_{j1} - \delta_{i1}\delta_{j0}). \qquad (2.20)$$

In matrix notation this is

$$m_{01} - m_{10}^{\dagger} = \tfrac{1}{2}I, \qquad (2.21)$$

where I is the $n_C \times n_C$ unit matrix. The matrices m_{00} and m_{11} are Hermitian (real and symmetric). When substituted into Eq. (2.18), Eq. (2.21) gives

$$\delta\Xi = \delta\alpha^{\dagger}m\alpha + (m\alpha)^{\dagger}\delta\alpha + \tfrac{1}{2}(\alpha_0^{\dagger}\delta\alpha_1 - \alpha_1^{\dagger}\delta\alpha_0). \qquad (2.22)$$

In the limit of completeness of the orbital basis for the quadratically integrable part of the open-channel orbitals, Eqs. (2.14) of the matrix variational method become equivalent to the close-coupling equations. A variational expression analogous to Eq. (2.22) can be derived for Ξ considered as a functional of the channel orbitals ψ_p or of the radial functions $f_p(r)$.

Feshbach partitioning, as defined by Eqs. (1.9) and (1.10), induces a corresponding partitioning in the linear space of radial functions. This is expressed by the condition that the radial channel orbitals $f_p(r)$ should be orthogonal to all radial basis orbitals $\eta_{pa}(r)$. The radial operator m^{pq} derived from the matrix orbital operator $\tilde{m}^{pq}$ of Eq. (1.15) is defined as an operator in the P space of the Feshbach partitioning. Hence m^{pq} acts only on radial channel orbital functions. The Lagrange multiplier terms in the generalized close-coupling Eq. (1.38) express the function obtained by the action of m^{pq} on any $f_q(r)$ as a linear combination of the Q-space basis radial functions. Hence, the function $\sum_q m^{pq}f_q$ vanishes in the P space, in complete analogy to the homogeneous matrix Eq. (2.14). The matrix m_{ij}^{pq} is just the matrix representation of the close-coupling operator m^{pq} in the open-channel orbital basis of the radial functions $F_{ip}(r)$.

Starting from Eq. (2.7) as definition of the variational functional, Feshbach partitioning leads through Eqs. (1.12) and (1.15) to the expression in terms of radial functions,

$$\Xi_{st} = \sum_p \sum_q \int_0^\infty f_{ps}(r) m^{pq} f_{qt}(r) \, dr, \tag{2.23}$$

or in matrix notation

$$\Xi = f^\dagger m f. \tag{2.24}$$

An infinitesimal variation of f that maintains Eq. (1.5) induces a variation of Ξ,

$$\delta\Xi = \delta f^\dagger m f + (mf)^\dagger \delta f + f^\dagger (m - m^\dagger)\delta f. \tag{2.25}$$

The last term here is a formal notation for a surface integral, to be evaluated in the limit of large r. Since it depends only on the asymptotic behavior of the channel orbitals, it is identical with the last term of Eq. (2.18) or of Eq. (2.22) and can be expressed in terms of the coefficients α_{ipq}. Hence

$$\delta\Xi = \delta f^\dagger m f + (mf)^\dagger \delta f + \tfrac{1}{2}(\alpha_0^\dagger \delta\alpha_1 - \alpha_1^\dagger \delta\alpha_0). \tag{2.26}$$

2.2. The Hulthén–Kohn Variational Principle and Variational Bounds

A variational expression such as Eq. (2.26) can be used and interpreted in several different ways. In particular, if f is a solution of the close-coupling equations, then mf is orthogonal to δf (restricted to the orbital P space by the orthogonality conditions), and Eq. (2.26) reduces to

$$\delta\Xi = \tfrac{1}{2}(\alpha_0^\dagger \delta\alpha_1 - \alpha_1^\dagger \delta\alpha_0). \tag{2.27}$$

This equation, which is characteristic of scattering theory, shows that Ξ is not, in general, stationary even for an exact solution of the Schrödinger equation because of the surface integral terms arising from open scattering channels. However, combining these terms with Ξ, Eq. (2.27) leads to various multichannel versions of the variational principle originally derived by Hulthén (1944, 1948), Kohn (1948), and Kato (1950).

As applied to the K matrix, Kohn's variational principle can be derived by restricting variations to those that maintain the canonical form

$$\alpha_0 = I, \qquad \alpha_1 = K, \tag{2.28}$$

Then

$$\delta\alpha_0 = 0, \qquad \delta\alpha_1 = \delta K, \tag{2.29}$$

and Eq. (2.27) becomes

$$\delta\Xi = \tfrac{1}{2}\delta K, \tag{2.30}$$

or

$$\delta[K] = 0, \tag{2.31}$$

where $[K]$ is the Kohn functional

$$[K] = K - 2\Xi. \tag{2.32}$$

Since Ξ vanishes for an exact scattering solution, the Kohn functional provides an expression for the K matrix that is stationary with respect to variations about an exact solution of the close-coupling equations.

In the matrix variational method, the matrix equations (2.17) do not in general have a solution. In this case, variational principles are used to select the best possible values of the free coefficients α. For variations restricted by Eqs. (2.28) and (2.29), Eq. (2.22) becomes

$$\delta\Xi = \delta K^\dagger(m_{10} + m_{11}K_t) + (m_{10} + m_{11}K_t)^\dagger \delta K + \tfrac{1}{2}\delta K. \tag{2.33}$$

Here K_t is an estimated trial K matrix, which cannot be assumed to be exact, and δK is an infinitesimal variation about K_t. The variational principle is applied in two steps. In the first, K_t is chosen to make $[K]$ stationary within the manifold of variations of K_t. This requires

$$m_{10} + m_{11}K_t = 0 \tag{2.34}$$

or

$$K_t = -m_{11}^{-1}m_{10}. \tag{2.35}$$

In the second step, the K matrix is estimated by the stationary value of the Kohn functional,

$$[K] = K_t - 2(m_{00} + m_{01}K_t + K_t^\dagger m_{10} + K_t^\dagger m_{11}K_t), \tag{2.36}$$

from Eqs. (2.16), (2.28), and (2.32). On substituting Eqs. (2.34) and (2.35) and using Eq. (2.21) to eliminate m_{01}, Eq. (2.36) becomes the Kohn formula,

$$[K] = -2(m_{00} - m_{10}^\dagger m_{11}^{-1}m_{10}). \tag{2.37}$$

Since m_{00} and m_{11} can be taken to be real and symmetric, Eq. (2.37) is obviously a real symmetric matrix. From its derivation, Eq. (2.37) remains valid in the limit of an exact scattering solution. This proves the symmetry of the exact K matrix. The obvious difficulty with this formula is that there is no guarantee that the matrix m_{11} us not singular (Nesbet, 1968, 1969a). At a zero eigenvalue of m_{11}, K_t or $m_{11}^{-1}m_{10}$ is not defined unless the columns of the matrix m_{10} are orthogonal to the corresponding eigenvector. Specific examples show that m_{10} does not have this property in general, except in the limit of an exact solution of the scattering equations, and that the number of zero eigenvalues of m_{11} increases with the number of basis functions used to represent the component Ψ_Q of the scattering wave function. The resulting anomalous singularities in $[K]$ will be discussed in detail in Section 2.3.

Analogous formulas are obtained for the matrix K^{-1} by restricting variations to the alternative canonical form

$$\alpha_0 = K^{-1}, \qquad \alpha_1 = I, \tag{2.38}$$

which, like Eqs. (2.28), gives the reactance matrix K as $\alpha_1 \alpha_0^{-1}$. Then

$$\delta\alpha_0 = \delta K^{-1}, \qquad \delta\alpha_1 = 0, \tag{2.39}$$

and Eq. (2.27) becomes

$$\delta\Xi = -\tfrac{1}{2}\delta K^{-1} \tag{2.40}$$

or

$$\delta[K^{-1}] = \delta(K^{-1} + 2\Xi) = 0. \tag{2.41}$$

In this case Eq. (2.22), for the matrix variational method, becomes

$$\delta\Xi = \delta K^{-1\dagger}(m_{00}K_t^{-1} + m_{01}) + (m_{00}K_t^{-1} + m_{01})^{\dagger}\delta K^{-1} - \tfrac{1}{2}\delta K^{-1}. \tag{2.42}$$

If K_t^{-1} is chosen so that

$$m_{00}K_t^{-1} + m_{01} = 0 \tag{2.43}$$

or

$$K_t^{-1} = -m_{00}^{-1}m_{01}, \tag{2.44}$$

then the stationary value of the functional is

$$\begin{aligned}
[K^{-1}] &= K_t^{-1} + 2\Xi_t \\
&= 2(m_{11} - m_{01}^{\dagger}m_{00}^{-1}m_{01}).
\end{aligned} \tag{2.45}$$

This will be referred to as the *inverse Kohn* formula for the real symmetric functional $[K^{-1}]$, due originally to Hulthén (1948) and Rubinow (1955). In this case, anomalous singularities occur at zero eigenvalues of m_{00} (Nesbet, 1968, 1969a).

Several methods have been considered for single-channel scattering that cannot easily be generalized to multichannel scattering. For $n_C = 1$, K is just $\tan\eta$, where η is a phase shift, and the matrix m is a 2×2 unsymmetric matrix m_{ij}. Equations (2.17) for the two coefficients α_i are

$$\begin{aligned}
m_{00}\alpha_0 + m_{01}\alpha_1 &= 0 \\
m_{10}\alpha_0 + m_{11}\alpha_1 &= 0,
\end{aligned} \tag{2.46}$$

such that

$$\tan\eta = \alpha_1/\alpha_0. \tag{2.47}$$

The variational functional is

$$\Xi = m_{00}\alpha_0^2 + (m_{01} + m_{10})\alpha_0\alpha_1 + m_{11}\alpha_1^2. \tag{2.48}$$

Hulthén (1944) proposed to choose $\tan \eta_t$ so that Ξ should vanish. Then Eq. (2.31) implies that $\tan \eta_t$ is stationary. A particular difficulty with this method is that Eq. (2.48) gives a quadratic equation for $\tan \eta_t$. A unique choice of the double-valued root is implied by comparison with Kohn's formula (Demkov and Shepalenko, 1958), but the method fails when the roots are complex (Nesbet, 1968).

In general, the two equations (2.46) are inconsistent. However, if the determinant

$$\det m = m_{00}m_{11} - m_{01}m_{10} \tag{2.49}$$

should vanish, the equations would be consistent, and in fact Hulthén's condition $\Xi = 0$ would be satisfied. The Kohn and inverse Kohn formulas would give the same result. Malik (1962) and Rudge (1973) proposed including an extra parameter in the variational wave function that could be adjusted to made $\det m$ vanish. Calculations by Rudge (1975) of e^-–H phase shifts were carried out with this method. It successfully avoids the anomalous singularities of the Kohn method.

One striking property of the matrix elements m_{ij}^{pq} defined by Eq. (2.9) is that they all have common poles at zero eigenvalues $E_\alpha - E$ of the bound–bound matrix $M_{\mu\nu}$. The implications of this will be discussed in Section 2.3, below. In the single-channel case, this property can be exploited to give a simple formula (Harris, 1967), valid only at the poles E_α,

$$\tan \eta = -M_{\alpha,0}/M_{\alpha,1}, \tag{2.50}$$

which avoids the computation of free–free integrals. Here $M_{\alpha,i}$ is a bound–free integral referred to an eigenvector of $M_{\mu\nu}$ indexed by α. Equation (2.50) is derived by taking the limit of either Eq. (2.35) or (2.44) at $E = E_\alpha$. It can be shown (Nesbet, 1968) that the variationally corrected Kohn or inverse Kohn values of $[\tan \eta]$ differ from Eq. (2.50) and are well behaved at the poles E_α, which cancel out of the variational formulas.

Kato (1950) introduced an additional fixed phase parameter θ into the single-channel variational principle, so that Eqs. (2.1) become

$$F_{0p}^\theta(r) \sim k_p^{-1/2} \sin{(k_pr - \tfrac{1}{2}l_p\pi + \theta)},$$

$$F_{1p}^\theta(r) \sim k_p^{-1/2} \cos{(k_pr - \tfrac{1}{2}l_p\pi + \theta)}. \tag{2.51}$$

The variational derivation goes through as before, but now the Kohn and inverse Kohn formulas appear as special cases, with $\theta = 0$ and $\pi/2$, respectively, of a more general formula parameterized by θ,

$$[\tan{(\eta - \theta)}] = -2(m_{00}^\theta - m_{10}^\theta(m_{11}^\theta)^{-1}m_{10}^\theta). \tag{2.52}$$

Here

$$m^\theta = \begin{pmatrix} \cos\theta & \sin\theta \\ -\sin\theta & \cos\theta \end{pmatrix} \begin{pmatrix} m_{00} & m_{01} \\ m_{10} & m_{11} \end{pmatrix} \begin{pmatrix} \cos\theta & -\sin\theta \\ \sin\theta & \cos\theta \end{pmatrix}, \quad (2.53)$$

obtained from the orthogonal transformation of the matrix m implied by the change of basis functions indicated in Eq. (2.51). Possible criteria for the choice of θ will be discussed in Section 2.4.

The Kohn variational principle can be used to derive Eq. (2.6) for the coefficients defining the component Ψ_Q of the full-scattering wave function. The Kohn functional is given in matrix notation by Eq. (2.32),

$$[K] = K - 2\Xi, \quad (2.54)$$

where Ξ is now taken to be defined by Eq. (2.7) in terms of the full wave function Ψ. For infinitesimal variations of Ψ, if all matrix elements are real,

$$\delta\Xi = 2(\delta\Psi|H - E|\Psi) + \tfrac{1}{2}(\alpha_0^\dagger\delta\alpha_1 - \alpha_1^\dagger\delta\alpha_0), \quad (2.55)$$

where the final term, due to a surface integral that depends only on the general asymptotic form of Ψ expressed by Eq. (1.5), is the same as in Eqs. (2.22) and (2.26). If the coefficients α are restricted to the canonical form given by Eq. (2.28), this term reduces to $\tfrac{1}{2}\delta K$. It then follows from Eqs. (2.54) and (2.55) that

$$\delta[K] = -4(\delta\Psi|H - E|\Psi) \quad (2.56)$$

since the terms in δK cancel. Equation (2.56) shows that if $[K]$ is stationary for all variations, then Ψ is an exact wave function. It also shows that $[K]$ is stationary with respect to variations of each K-matrix element and with respect to any independent variable parameters used to define Ψ. Since the coefficients c_μ^{ip} in Eq. (2.4) are independent of the coefficients α_{ips} by construction and since $[K]$ depends on the coefficients c_μ^{ip} only through the functional Ξ, Eq. (2.6) follows immediately from this stationary property of $[K]$. Equation (2.37), the Kohn formula for the stationary value of $[K]$, follows from its stationary property with respect to variation of a trial K matrix.

For single-channel scattering, this argument obviously can be generalized directly to apply to the Kato functional of Eq. (2.52). The appropriate generalization for multichannel scattering requires further analysis, which will be presented in Section 2.4.

In the variational analysis of Kato (1951), the second-order terms are retained in Eq. (2.55) to give

$$\delta[K] = -4(\delta\Psi|H - E|\Psi) - 2(\delta\Psi|H - E|\delta\Psi). \quad (2.57)$$

If $\delta\Psi$ is a variation about an exact wave function, the first term on the right vanishes, leaving the second-order term. Under certain conditions, essentially that $\delta\Psi$ is orthogonal to all negative eigenvalue eigenstates of $H - E$ (which must include all open-channel states), $(\delta\Psi|H - E|\delta\Psi)$ is positive semidefinite. Hence $[K]$ provides lower bounds for individual diagonal elements of the K matrix, and the error matrix is negative semidefinite. Similarly, the inverse Kohn variational principle would provide upper bounds for elements of K^{-1}, or a positive semidefinite error matrix. This result is the basis for theories of variational bounds for eigenphases and K-matrix elements (Hahn *et al.*, 1962, 1963, 1964a,b; Sugar and Blankenbecler, 1964; Gailitis, 1965a; Hahn and Spruch, 1967). Such theories are most useful in the close-coupling method, when Eqs. (2.17) are solved exactly, but the operator m is approximated by a truncated closed-channel expansion. These theories establish rigorous bounds only when exact target atom wave functions are used in the close-coupling expansion, which in practice can only be done for hydrogen or one-electron ions, and then only if the open-channel equations are solved exactly. This severely limits their applicability in the present context. Hahn (1971) has formulated a "quasi-minimum" variational principle that relaxes the rigorous bound argument in order to allow variational approximation to the solutions of the open-channel equations. In practice, it is found that computed K-matrix elements or phase shifts show stationary behavior as functions of variational parameters in Ψ_Q when the continuum part of Ψ is represented to relatively high accuracy.

2.3. Anomalies in the Kohn Formalism

The anomalous singularities inherent in the Kohn formula were first considered by Schwartz (1961a,b), who encountered these anomalies in variational calculations of e^-–H elastic scattering. Nevertheless, the s-wave phase shifts obtained by Schwartz, after empirically smoothing the variationally computed numbers, still stand as the most accurate work available.

Figure 2.1 (Schwartz, 1961b, Fig. 1) shows the ^3S phase shift at fixed wave number $k = 0.4a_0^{-1}$, evaluated with wave functions expanded in basis sets containing from three to 34 functions of r_1, r_2, and r_{12}. Values of $-[\tan\eta]/k$ computed from the Kohn formula are shown for a range of values of a scale parameter κ. Although $[\tan\eta]$ is not bounded, the curves for different basis dimensions appear to converge in a limited range of κ that increases in length with the basis set dimension. A spurious pole is evident near $\kappa = 1.6$ in the 7×7 curve. Schwartz was able to extrapolate results of this kind to approximate convergence, assuming that the nested flat portions of the curves for different basis dimensions were exhibiting extremal

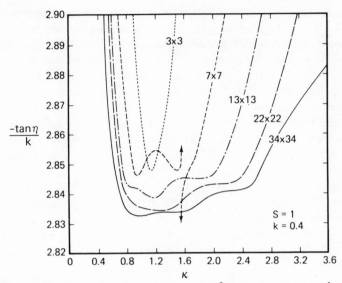

Figure 2.1. Variation of tan η/k with parameter κ, e^{-}–H ^3S scattering, $k = 0.4a_0^{-1}$ (Schwartz, 1961b, Fig. 1).

behavior and that other structure was spurious. The value of tan η/k deduced from these data was $-2.833(2)$ with error as indicated in the last digit, or $\eta = 2.2938(4)$ radians.

Figure 2.2 (Schwartz, 1961a, Fig. 2), shows the ^3S phase shift at $k = 0.8a_0^{-1}$, plotting tan η /k as a function of the scale parameter κ. Here the behavior is more complicated, showing numerous spurious poles and a region of values (for small κ) quite unrelated to the majority of data points. Because tan η is large, small deviations in the phase shift are magnified, as indicated by the spread of 0.003 radians shown in the figure. From these data Schwartz was able to estimate the value of tan η/k as $-17.2(6)$, corresponding to $\eta = 1.643(3)$ radians.

Brownstein and McKinley (1968) studied the anomalies in the Kohn method in variational calculations of scattering by a square-well potential, for which exact solutions are known. Figure 2.3 (Brownstein and McKinley, 1968, Fig. 2) shows values of tan η/k computed at fixed energy ($k = 0.8a_0^{-1}$, well depth -4 a.u. for $r \leq 1a_0$) for a range of values of a parameter κ from an exponential factor in the basis functions. Four quadratically integrable basis functions were used. Figure 2.3 clearly shows anomalous poles, and also illustrates the approximate stationary behavior utilized by Schwartz, since the more horizontal segments of the plotted curve approach the exact value from below. These anomalies in the Kohn formula were originally attributed to the effect of eigenvalues of the bound–bound matrix, causing singularities

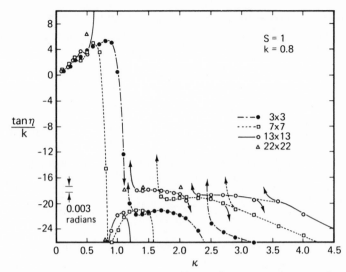

Figure 2.2. Variation of tan η/k with parameter κ, e^-–H ^3S scattering, $k = 0.8a_0^{-1}$ (Schwartz, 1961a, Fig. 2).

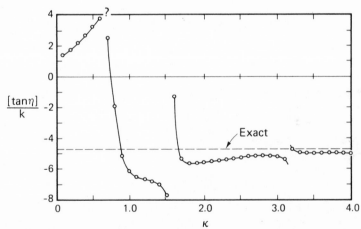

Figure 2.3. Variation of $[\tan \eta]/k$ with parameter κ, square well model, $k = 0.8a_0^{-1}$ (Brownstein and McKinley, 1968, Fig. 2).

in the effective optical potential through the energy denominator in Eq. (1.13). More detailed analysis, given below (Nesbet, 1968, 1969a), showed that the resulting poles exactly cancel out of either the Kohn or inverse Kohn formulas. The observed anomalies arise instead from the singular points of m_{11} or m_{00} in Eqs. (2.37) and (2.45), respectively. This fact is of great practical importance since singularities of the optical potential would be common to all variational methods and unavoidable, whereas in practice the singular points of m_{11} and m_{00} do not coincide. Hence anomalies can be avoided by a suitable choice of variational formalism.

Near an eigenvalue E_α of the bound–bound matrix $M_{\mu\nu}$, the behavior of matrix elements of m, given by Eq. (2.9), is

$$(E - E_\alpha)m_{ij}^{pq} = M_{ip,\alpha}M_{\alpha,jq} + O(E - E_\alpha), \tag{2.58}$$

where α indexes the eigenstate corresponding to E_α. Thus the matrix elements all have coincident simple poles as functions of energy. The residues define a dyadic matrix (of rank 1), constructed from column vectors $\mathbf{v}_i$ whose elements are $M_{\alpha,ip}$. Thus in matrix notation, suppressing open-channel indices, Eq. (2.58) is

$$(E - E_\alpha)m_{ij} = \mathbf{v}_i\mathbf{v}_j^\dagger + O(E - E_\alpha). \tag{2.59}$$

This dyadic structure leads to exact cancellation of the apparent singularities of the Kohn and inverse Kohn formulas at eigenvalues E_α (Nesbet, 1969a).

This can be shown most directly for single-channel scattering (Nesbet, 1968). The Kohn formula, Eq. (2.37), can be written as

$$[\tan \eta] = -(m_{10} + 2 \det m)/m_{11}, \tag{2.60}$$

where $\det m$ is given by Eq. (2.49). The inverse Kohn formula, Eq. (2.45), can be written as

$$[\cot \eta] = -(m_{01} - 2 \det m)/m_{00}. \tag{2.61}$$

From Eq. (2.59), for a single channel,

$$(E - E_\alpha) \det m = (E - E_\alpha)(m_{00}m_{11} - m_{01}m_{10}) \tag{2.62}$$

$$= (E - E_\alpha)^{-1}(v_0v_0v_1v_1 - v_0v_1v_1v_0) + \text{const} + O(E - E_\alpha) \tag{2.63}$$

$$= \text{const} + O(E - E_\alpha), \tag{2.64}$$

so that $\det m$ has only a simple pole at E_α. It follows immediately that in both Eqs. (2.60) and (2.61) numerator and denominator have coincident simple poles that cancel at E_α. Hence their behavior near E_α is given by

$$[\tan \eta] = -\frac{v_0}{v_1} - 2\frac{\text{const}}{v_1^2} + O(E - E_\alpha), \tag{2.65}$$

$$[\cot \eta] = -\frac{v_1}{v_0} + 2\frac{\text{const}}{v_0^2} + O(E - E_\alpha). \tag{2.66}$$

Since the matrix elements v_0 and v_1 vary slowly with energy, there are no spurious singularities in either formula at the eigenvalues E_α. In both cases the limiting value differs from Eq. (2.50), the Harris formula (Nesbet, 1968; Morawitz, 1970).

In the multichannel case, the Kohn formula, written out as a matrix product, is

$$[K]_{pq} = -2\left(m_{00}^{pq} - \sum_s \sum_t m_{10}^{sp}(m_{11}^{-1})^{st} m_{10}^{tq} \right). \tag{2.67}$$

It can easily be verified, by expanding the numerator of the following expression in minors, that the Kohn formula can be expressed as the ratio of two determinants,

$$[K]_{pq} = \frac{-2}{|m_{11}^{st}|} \begin{vmatrix} m_{00}^{pq} & \cdots & m_{10}^{tq} \\ \vdots & \ddots & \\ m_{10}^{sp} & & m_{11}^{st} \end{vmatrix}. \tag{2.68}$$

The $(n_C + 1)$-order determinant in the numerator is constructed by adding the indicated first row and column to the matrix m_{11}, whose determinant is in the denominator. In order to examine behavior near any eigenvalue E_α, common factors in rows and columns can be removed by defining, for each element of either matrix,

$$f_{ij}^{pq}(E) = (E - E_\alpha) m_{ij}^{pq} / v_i^p v_j^q, \tag{2.69}$$

where the elements v_i^p are bound–free integrals as in Eq. (2.59). As E approaches E_α,

$$f_{ij}^{pq}(E) = 1 + O(E - E_\alpha), \tag{2.70}$$

assuming that none of the v_i^p vanish. Then, canceling common factors in Eq. (2.68),

$$[K]_{pq} = \frac{-2v_0^p v_0^q}{(E - E_\alpha)|f_{11}^{st}|} \begin{vmatrix} f_{00}^{pq} & \cdots & f_{10}^{tq} \\ \vdots & \ddots & \\ f_{10}^{sp} & & f_{11}^{st} \end{vmatrix}. \tag{2.71}$$

At E_α, both determinants here are of rank 1 since all their elements are unity. A theorem of matrix theory (Frazer *et al.*, 1947, p. 17) states that if $\Delta(\lambda)$ is the determinant of a matrix of order n, whose elements are functions of λ, and if for some λ_0 the matrix becomes singular, of rank r, then $(\lambda - \lambda_0)^{n-r}$ is a factor of $\Delta(\lambda)$. It follows immediately from this theorem, for any matrix of order n whose elements are functions $f_{ij}^{pq}(E)$, since such a matrix reduces to rank 1 at E_α, that $(E - E_\alpha)^{n-1}$ is a factor of its determinant. Since the numerator in Eq. (2.71) is of order $n_C + 1$, while the denominator is of order n_C but with an additional factor $(E - E_\alpha)$, these factors cancel, and $[K]_{pq}$ has no singularity at E_α. A similar theorem can be proved for the multichannel inverse Kohn formula.

Although the Kohn and inverse Kohn formulas are well behaved at the bound–bound energy eigenvalues E_α, they still have spurious poles at the zero eigenvalues of the matrices m_{11} and m_{00}, respectively. This is shown clearly by model calculations (Nesbet, 1968, 1969a; Truhlar *et al.*, 1974; Callaway, 1978). In contrast to the weak energy dependence of the bound–free matrix elements in Eqs. (2.58) and (2.59), the poles E_α impose a strong energy dependence on the matrix elements m_{ij}^{pq}. In particular, by differentiating Eq. (2.59) and neglecting the weak energy dependence, near a pole E_α,

$$\frac{\partial m_{ij}}{\partial E} \cong -\frac{\mathbf{v}_i \mathbf{v}_j^\dagger}{(E - E_\alpha)^2}. \qquad (2.72)$$

For m_{00} and m_{11}, the numerator is positive definite, so the derivative matrix is approximately negative definite. The pattern imposed on both m_{00} and m_{11} is that one eigenvalue descends from $+\infty$ to $-\infty$ going from one pole E_α to the next. Hence the effect of this strong energy dependence is to produce at least one zero eigenvalue in each case, between the adjacent poles. There is nothing in the theory to make the singular points of m_{00} and m_{11} coincide.

When an additional parameter, such as the scaling constant κ in the work of Schwartz (1961a,b), is introduced in the variational wave function, the resulting anomalies are functions of both E and κ and appear as functions of κ if E is fixed. This is the case in the examples shown in Figs. 2.1, 2.2, and 2.3. An example of typical energy dependence of matrix elements m_{10} and m_{11} near an anomaly in the Kohn formula is shown in Fig. 2.4

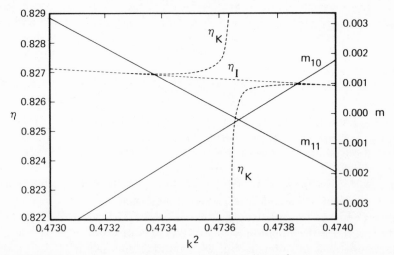

Figure 2.4. Variation of m_{10}, m_{11}, η_K, and η_I with energy, $e^- - H^1 S$ scattering. Phase shifts modulo π radians (Callaway, 1978b, Fig. 2).

(Callaway, 1978b, Fig. 2). The example is taken from a three-state algebraic close-coupling calculation of the $e^- - H$ ^1S phase shift. As a function of energy (k^2 here) m_{11} decreases monotonically, and $[\tan \eta_K]$ has a narrow region of anomaly centered at the point where m_{11} passes through zero. This causes a local increase of η_K through π radians, behavior described as a *pseudoresonance*. In this example $[\cot \eta_I]$, the inverse Kohn formula, has no nearby singularity ($m_{00} \neq 0$), and η_I varies smoothly. At a true resonance, η_K and η_I would remain close to each other, both showing an increase through π radians.

As the number of basis functions for the bound–bound matrix is increased, the number of eigenvalues E_α increases, as does the number of points where m_{00} and m_{11} are singular. Although the number of anomalous singularities in either the Kohn or inverse Kohn formula grows without limit, it is found in practice that their apparent width decreases at least as rapidly as the spacing between them (Brownstein and McKinley, 1968; Nuttall, 1969).

The effective width of these anomalies, as functions of energy, can be understood in terms of an argument given by Harris and Michels (1971). For single-channel scattering, the Kohn formula is

$$[\tan \eta] = -2(m_{00} - m_{10}^2/m_{11}). \qquad (2.73)$$

A pole occurs at $m_{11} = 0$ unless m_{10} also vanishes for the same value of E. For an exact solution, det m as given by Eq. (2.49) vanishes for all energies, so in this case

$$m_{00}m_{11} = m_{01}m_{10}, \qquad (2.74)$$

which implies that either m_{01} or m_{10} must go to zero with m_{11}. The typical situation in variational calculations is that m_{10} and det m are small near such zeroes, and because of Eq. (2.21), m_{01} is approximately equal to $+\frac{1}{2}$. Equations (2.72) and (2.74) show that m_{10}, like m_{11}, will tend to have small values between poles at eigenvalues E_α, but the slope may be either positive or negative. Thus m_{10} and m_{11} may both have zeroes near the middle of the interval between adjacent values E_α, but these zeroes coincide only when det m vanishes. In general, the spacing between these zeroes decreases as the eigenvalues E_α become more dense, and this spacing controls the effective width of the anomaly due to the zero of m_{11}. Figure 2.4 illustrates this behavior.

From Eq. (2.73), if m_{11} and m_{10} have coincident zeroes, the values of $[\tan \eta]$ is just $-2m_{00}$. Because these zeroes are separated, a fluctuation from this value takes place when $|m_{10}^2/m_{00}m_{11}|$ is not small compared with unity. If both m_{00} and det m are different from zero, the local value of m_{10} (with $m_{01} = \frac{1}{2}$ and $m_{11} = 0$) is

$$m_{10} = \frac{\det m}{-m_{01}} = -2 \det m. \qquad (2.75)$$

Then the range of fluctuation is given by

$$|m_{11}| \lesssim |4(\det m)^2/m_{00}|. \tag{2.76}$$

This decreases quadratically with det m, which vanishes for an exact wave function. A similar argument can be used to estimate the width of the anomaly in $[\cot \eta]$ as a function of E.

2.4. Anomaly-Free Methods

The work of Schwartz (1961a,b) illustrates the difficulty of using the original Kohn formula for accurate calculations. Many individual calculations must be carried out in order to establish a meaningful value of a single phase shift at a given energy. In order to make accurate calculations feasible for general applications, it is necessary to find a method free of such computational artifacts as spurious singularities.

The strong energy dependence of the matrix elements m_{ij}^{pq} near bound–bound eigenvalues E_α causes them to have coincident poles and causes the submatrices m_{00} and m_{11} to have zero eigenvalues between these poles. Since these zeroes occur in between poles, their exact location is not forced. In general the zero eigenvalues of m_{00} and m_{11} occur at different energies, so that the resulting spurious singularities in the inverse Kohn and Kohn formulas do not coincide. This observation is the basis of the *anomaly-free* (AF) method (Nesbet 1968, 1969a), which consists simply of using alternatively the Kohn or inverse Kohn formula if the ratio of determinants $|m_{11}|/|m_{00}|$ is greater than or less than unity, respectively.

In practice, the AF method is not completely satisfactory because it can introduce a noticeable discontinuity in regions of strong energy dependence and because sometimes zero eigenvalues of one of m_{11} and m_{00} fall within the range of anomaly of the other as estimated by Eq. (2.76). The ideal method, which should be viable for multichannel scattering, would obtain a real symmetric K matrix without spurious singularities, optimized in the sense of being stationary with respect to variations of free parameters and varying continuously as a function either of energy or of variational parameters.

In the case of single-channel scattering, two anomaly-free methods based on the Kohn formalism had actually been proposed prior to the analysis, given above, that properly identified the anomalies with zeroes of m_{11} and m_{00}.

The formula of Kato (1950), Eq. (2.52), can be used to avoid spurious singularities by choice of the free parameter θ (Takatsuka and Fueno, 1979a, b). Any criterion which ensured that m_{11}^{θ} would be outside the range of anomalous behavior indicated in Eq. (2.76) would define an anomaly-free method. An obvious choice is to maximize the magnitude of m_{11}^{θ} with respect to variations of θ. The multichannel generalization of this criterion is used in the *interpolated anomaly-free* (IAF) method (Nesbet, 1978), to be described below. This criterion in general ensures that the method is superior to either the Kohn or inverse Kohn formula and also that results vary continuously with respect to energy or other parameters.

Another single-channel method that successfully avoids anomalies was proposed by Malik (1962), before the nature of the anomalies was understood, and more recently by Rudge (1973), who emphasized its anomaly-free aspect. This method uses a variable parameter in the variational wave function. This parameter is adjusted to make det m vanish. This ensures that zeroes of m_{10} and m_{11} coincide, so anomalies are avoided. For multichannel scattering, the corresponding condition requires n_C simultaneous zero eigenvalues of the m matrix, so n_C conditions are required, and the method cannot be generalized.

The method of Kato (1950) introduces an orthogonal transformation of the m matrix, as indicated in Eq. (2.53). Harris and Michels (1969a, 1971) proposed the use of a general $2n_C \times 2n_C$ orthogonal transformation in the multichannel case, chosen to minimize the norm of $m\alpha$ as defined by Eqs. (2.14) and (2.17). This is the *minimum norm* (MN) method. Following analysis and notation of Nesbet and Oberoi (1972), the MN method determines a trial matrix α_t by minimizing the trace of the quadratic form

$$(m\alpha)^{\dagger}m\alpha = \alpha^{\dagger}m^{\dagger}m\alpha. \tag{2.77}$$

Then q_t is the $2n_C \times n_C$ matrix of those column eigenvectors of the positive definite real symmetric matrix $m^{\dagger}m$ that correspond to the n_C smallest eigenvalues.

Unlike the Hulthén–Kohn formalism, the MN method does not provide a unique prescription for refining the trial matrix α_t obtained by diagonalizing $m^{\dagger}m$. Harris and Michels (1969a) proposed use of the Kohn functional $[K]$ of Eq. (2.32) in the form

$$[K] = K_t - 2(m_{00} + m_{01}K_t + K_t^{\dagger}m_{10} + K_t^{\dagger}m_{11}K_t) \tag{2.78}$$

with

$$K_t = \alpha_{1t}\alpha_{0t}^{-1}, \tag{2.79}$$

as required by Eq. (1.17). Alternatively, the inverse Kohn functional could be used. In fact, the use of the Kohn functional with K_t estimated by Eq. (2.79) for the minimum-norm α_t is not really justified since $[K]$ should be made stationary with respect to variation of the elements of K_t, leading back to the Kohn formula. Nevertheless, defining a trial matrix α_t by minimizing the norm of $m\alpha$, which should vanish for an exact wave function, is an intuitively reasonable procedure. The problem that remained unresolved was to define a variational functional $[\alpha]$ that is made stationary by this choice of α_t.

The central idea of the minimum norm method and of Kato's method is to introduce a preliminary orthogonal transformation of the asymptotic wave functions and hence of the m matrix so as to reduce the error of the subsequent optimizing step in which a stationary variational functional is evaluated. Despite the relative success of several alternative methods, to be discussed below, the recently developed IAF method (Nesbet, 1978) is the only one making use of a preliminary transformation and a stationary functional that is free of anomalies, continuous, and produces a symmetric K matrix.

The relationship given by Eq. (2.21),

$$m_{01} - m_{10}^{\dagger} = \tfrac{1}{2}I, \tag{2.80}$$

reduces the surface integral in $\delta \Xi$ to the form given by Eq. (2.27), which is valid for an exact wave function. This form is used to derive both the Kohn and inverse Kohn formulas for stationary functionals.

In the Kato method, for single-channel scattering, the 2×2 transformation of m_{ij} parameterized by the phase angle θ, Eq. (2.53), preserves Eq. (2.80). This is no longer true for the general $2n_C \times 2n_C$ orthogonal transformation considered in the minimum norm method. The essential idea of the IAF method (Nesbet, 1978b) is to consider the subclass of orthogonal transformations that preserve Eq. (2.80). Then the Kohn formalism defines a stationary functional in the space of transformed asymptotic wave functions.

The full $2n_C \times 2n_C$ transformation matrix to be considered is of the form

$$u = (\alpha\beta), \tag{2.81}$$

where α and β are both $2n_C \times n_C$ matrices, each consisting of n_C column vectors. For an orthogonal transformation, the orthonormality conditions are

$$\alpha^{\dagger}\alpha = \beta^{\dagger}\beta = I,$$
$$\alpha^{\dagger}\beta = \beta^{\dagger}\alpha = 0. \tag{2.82}$$

The transformed m matrix is

$$m' = u^{\dagger}mu$$

$$= \begin{bmatrix} m'_{00} & m'_{11} \\ m'_{10} & m'_{11} \end{bmatrix}$$

$$= \begin{bmatrix} \alpha^\dagger m\alpha & \alpha^\dagger m\beta \\ \beta^\dagger m\alpha & \beta^\dagger m\beta \end{bmatrix}. \tag{2.83}$$

From Eq. (2.80), the antisymmetric part of the matrix m is $\frac{1}{4}J$, where

$$J = \begin{bmatrix} 0 & I \\ -I & 0 \end{bmatrix}, \tag{2.84}$$

if I is the $n_C \times n_C$ unit matrix. Equation (2.80) is preserved by the group of orthogonal transformations whose matrices u commute with J. For these transformations the antisymmetric part of m' is the same as that of m, defined by Eq. (2.80). It will be shown here that these matrices are of the form

$$u = \begin{bmatrix} CX & -SX \\ SX & CX \end{bmatrix}, \tag{2.85}$$

where X is an arbitrary orthogonal matrix, and C and S are real and symmetric $n_C \times n_C$ matrices such that

$$C^2 + S^2 = I, \qquad CS = SC. \tag{2.86}$$

A matrix of the form of Eq. (2.85) defines a real symmetric matrix

$$T = SC^{-1} = \tan \Delta \tag{2.87}$$

where Δ is a real symmetric phase matrix such that

$$C = \cos \Delta, \qquad S = \sin \Delta. \tag{2.88}$$

In terms of the submatrices α_i, β_i of u defined by Eq. (2.81), Eq. (2.87) is

$$T = \alpha_1 \alpha_0^{-1} = -\beta_0 \beta_1^{-1}. \tag{2.89}$$

This is a real symmetric matrix only if u has the special form considered here. Expressed in block form, the general $2n_C \times 2n_C$ orthogonal matrix is

$$u = \begin{bmatrix} \alpha_0 & \beta_0 \\ \alpha_1 & \beta_1 \end{bmatrix}, \tag{2.90}$$

where

$$\alpha_0^\dagger \alpha_0 + \alpha_1^\dagger \alpha_1 = I$$
$$\beta_0^\dagger \beta_0 + \beta_1^\dagger \beta_1 = I \tag{2.91}$$
$$\alpha_0^\dagger \beta_0 + \alpha_1^\dagger \beta_1 = 0.$$

To simplify the argument here, it will be assumed that α_0 and β_1 are not singular. Then Eq. (2.91) implies that

$$\beta_0 \beta_1^{-1} = -(\alpha_1 \alpha_0^{-1})^\dagger. \tag{2.92}$$

Define the matrix $T = \alpha_1 \alpha_0^{-1}$. Then if u is of the special form

$$u = \begin{bmatrix} CX & -SY \\ SX & CY \end{bmatrix},$$ (2.93)

where C and S are real symmetric matrices as given by Eq. (2.86) or (2.88) and X and Y are orthogonal matrices, it follows that

$$T = SX(CX)^{-1} = SC^{-1} = \tan \Delta.$$ (2.94)

Hence

$$T = T^\dagger.$$ (2.95)

Conversely, if $T = \alpha_1 \alpha_0^{-1}$ is real and symmetric it implies Eq. (2.93). This can be shown as follows. Equation (2.95) defines $\tan \Delta$ and hence, by Eqs. (2.88), matrices C and S with the required properties. Given

$$\alpha_1 \alpha_0^{-1} = T = SC^{-1}$$ (2.96)

then C is nonsingular, and

$$\alpha_1 = SC^{-1} \alpha_0 = SX,$$ (2.97)

where X is the nonsingular matrix

$$X = C^{-1} \alpha_0.$$ (2.98)

From Eq. (2.92),

$$\beta_0 \beta_1^{-1} = -T^\dagger = -T = -SC^{-1}.$$ (2.99)

Then

$$\beta_0 = -SC^{-1} \beta_1 = -SY,$$ (2.100)

where Y is the nonsingular matrix

$$Y = C^{-1} \beta_1.$$ (2.101)

From Eq. (2.98), α_0 is CX, and from Eq. (2.101), β_1 is CY. Hence u has the structure indicated in Eq. (2.93). For an orthogonal transformation, the nonsingular matrices X and Y must in fact be orthogonal matrices.

It can now be shown that an orthogonal matrix u commutes with matrix J of Eq. (2.84) if and only if it has the form given in Eq. (2.85). It is immediately obvious that any matrix of the form Eq. (2.85) commutes with J, as a consequence of the properties of matrices C and S given in Eqs. (2.86). Conversely, if u is given by Eq. (2.90), the condition

$$uJ = Ju$$ (2.102)

requires

$$\beta_0 = -\alpha_1, \qquad \beta_1 = \alpha_0. \tag{2.103}$$

Then Eq. (2.91) implies

$$-\alpha_0^\dagger \alpha_1 + \alpha_1^\dagger \alpha_0 = 0 \tag{2.104}$$

or

$$\alpha_1 \alpha_0^{-1} = (\alpha_0^\dagger)^{-1} \alpha_1^\dagger = (\alpha_1 \alpha_0^{-1})^\dagger. \tag{2.105}$$

Hence Eq. (2.95) is satisfied, which was shown above to imply that u is of the form given by Eq. (2.93). Since Eq. (2.103) requires $Y = X$, this implies Eq. (2.85). This establishes a canonical form for orthogonal matrices that commute with matrix J and hence preserve Eq. (2.80) for the asymptotic normalization of scattering wave functions. These matrices form a group that can be called the *physically relevant* group since they preserve a property of an exact wave function.

The IAF method (Nesbet, 1978b) is defined by choosing a preliminary orthogonal transformation of the physically relevant group so as to maximize the determinant $|m'_{11}|$. The Kohn variational formalism is used in the transformed linear space of asymptotic channel orbitals. In this transformed space, the most general optimizing transformation is of the form

$$[\alpha] = \alpha + \beta K', \tag{2.106}$$

where α and β are the rectangular matrices defined by Eq. (2.81), and K' is an $n_C \times n_C$ matrix. The K matrix, defined by

$$K = \alpha_1 \alpha_0^{-1} \tag{2.107}$$

is not affected by any nonsingular transformation that does not mix the column vectors of α and β. In Eq. (2.106) matrices α and β are fixed and only K' varies. For an infinitesimal variation $\delta K'$, the variation of α, from Eq. (2.106), is

$$\delta\alpha = \beta\delta K', \tag{2.108}$$

and from Eq. (2.18) the variation of the functional Ξ is

$$\delta\Xi = \delta K'^\dagger(m'_{10} + m'_{11}K') + (m'^\dagger_{10} + K'^\dagger m'_{11})\delta K' + (m'_{01} - m'^\dagger_{10})\delta K', \tag{2.109}$$

where the matrices m'_{ij} are defined by Eqs. (2.83). For transformations of the physically relevant group,

$$m'_{01} - m'^\dagger_{10} = \tfrac{1}{2}I \tag{2.110}$$

so that the analysis leading to the Kohn formula, Eq. (2.37), can be followed in the transformed system. Hence the functional

$$[K'] = -2(m'_{00} - m'^{\dagger}_{10}(m'_{11})^{-1}m'_{10}) \tag{2.111}$$

is stationary and equals K'_t for an exact wave function. From Eqs. (2.106) and (2.107) the K matrix is

$$K = (\alpha_1 + \beta_1[K'])(\alpha_0 + \beta_0[K'])^{-1}. \tag{2.112}$$

The purpose of the IAF condition, maximizing the determinant $|m'_{11}|$, is to avoid the zeroes of $|m'_{11}|$, which would cause anomalous poles in Eq. (2.111). Since, from Eq. (2.83),

$$m'_{11} = \beta^{\dagger}m\beta$$
$$= X^{\dagger}(uX^{\dagger})^{\dagger}m(uX^{\dagger})X, \tag{2.113}$$

where

$$uX^{\dagger} = \begin{bmatrix} C & -S \\ S & C \end{bmatrix}, \tag{2.114}$$

$|m'_{11}|$ does not depend on X, so the reduced canonical form $uX^{\dagger}$ is all that needs to be considered. This is completely defined by the matrix T of Eq. (2.87). Assuming this reduced form of the preliminary transformation, Eq. (2.112) becomes

$$K = (S + C[K'])(C - S[K'])^{-1}, \tag{2.115}$$

a result originally derived by Seaton (1966). As given by Eq. (2.111), $[K']$ is a symmetric matrix. If $[K']$ is symmetric, it can be proven that K given by Eq. (2.115) is symmetric (Nesbet and Oberoi, 1972).

Since no direct algorithm is known for constructing u so as to maximize $|m'_{11}|$, an iterative procedure is required. It is found in exploratory calculations that restricting C and S to be diagonal matrices produces satisfactory results, so more general transformations are probably not necessary. This restricted IAF method (RIAF) obtains C and S in the form

$$C_{pq} = \delta_{pq} \cos \theta_p,$$
$$S_{pq} = \delta_{pq} \sin \theta_p, \tag{2.116}$$

where the n_C phase angles θ_p are chosen to maximize $|\det m'_{11}|$. The Kohn method corresponds to taking all $\theta_p = 0$, while for the inverse Kohn method all $\theta_p = \frac{1}{2}\pi$. The RIAF formalism is the most direct multichannel generalization of the use of Kato's functional, Eq. (2.52), with θ determined by maximizing $|m_{11}|$. The RIAF method clearly ensures that $|\det m'_{11}|$ is greater than its value in either the Kohn or inverse Kohn limit, and hence should be considerably more effective in avoiding spurious anomalies than the AF method, which makes a choice of these two limiting cases. The effectiveness of this method is verified by model calculations (Nesbet, 1978b).

In attempting to avoid spurious singularities in Eq. (2.111), an alternative strategy is to force m'_{10} to vanish. Any real square matrix can be converted to upper triangular form by some unitary transformation, or by an orthogonal transformation if all eigenvalues are real. Such a transformation is used in the *optimized anomaly-free* (OAF) method (Nesbet and Oberoi, 1972). This method has been used extensively for electron–atom scattering calculations. The OAF method has several difficulties that make it less satisfactory than the RIAF method described above. In general, Eq. (2.110) is not valid for transformations that cause m'_{10} to vanish. This modifies Eq. (2.111) so that $[K']$ and K are not exactly symmetric. In practice, the computed K matrix must be symmetrized before computing cross sections. Special problems occur when m (which is a real unsymmetric matrix) has complex eigenvalues. A transformation to reorder diagonal elements must be used to ensure that m'_{01} remains nonsingular in the limit of an exact scattering solution. These properties of the method introduce some arbitrariness and lead in some applications to irregular energy dependence that appears to be a computational artifact. A special algorithm for reordering the eigenvalues of m' in the OAF method has been published (Nesbet, 1978b) together with a modified method (OAF2) that obtains a symmetric K matrix. Both OAF and OAF2 methods exhibit minor discontinuities in results as a consequence of the necessary reordering of eigenvalues of m.

The present analysis, with the variational principle expressed in terms of Eq. (2.109), was applied to the preliminary transformation of the minimum-norm method by Nesbet and Oberoi (1972) to define an optimized (minimum-norm) method (OMN). In this method, a matrix T as in Eq. (2.96) is defined as the symmetric part of $\alpha_1 \alpha_0^{-1}$. Then Eqs. (2.111) and (2.115) can be used. Since m'_{11} can still be singular, as shown by trial computations, the OMN method does not avoid spurious singularities.

An alternative method proposed by Wladawsky (1973) is called the *variational least-squares* (VLS) method. In fact the VLS method follows the original Kohn method in truncating the derived overdetermined system of equations (Abdallah and Truhlar, 1974) rather than using the true least-squares procedure provided by the generalized inverse (Penrose, 1955; Householder, 1964, pp. 8–10, 28). The VLS method differs from all other methods considered here by not partitioning the quadratically integrable variational basis (of generally large dimension N_Q) from the asymptotic open-channel basis (of dimension $2n_C$). Since $2n_C \ll N_Q$ in practical applications, the VLS method requires manipulations of very large matrices. This is avoided in the methods considered here once m has been constructed, which can be done using an efficient algorithm (Nesbet, 1971). In comparison, the VLS method is probably less efficient for practical applications unless specialized new matrix algorithms can be developed. Another version of the variational least-squares method has been used by Abdel-Raouf and Belschner (1978) for accurate calculations of $e^{\pm}$–H s-wave phase shifts. These calculations show smooth convergence as the basis set is increased. Abdel-Raouf (1979) has extended these calculations to a study of upper and lower bounds of the s-wave phase shifts.

2.5. Variational R-Matrix Method

The derivative matrix or R matrix (Wigner and Eisenbud, 1947; Lane and Thomas, 1958; Breit, 1959) is defined by the relationship between radial channel orbitals $f_{ps}(r)$ and their derivatives evaluated at some r_0. Thus

$$f_{ps}(r_0) = \sum_q R_{pq} r_0 f'_{qs}(r_0), \tag{2.117}$$

where R_{pq} is the R matrix. As defined here, it is dimensionless and can be shown to be a real symmetric matrix for solutions of close-coupling equations. The indices refer to both closed and open channels, which are not distinguished at a finite channel radius r_0.

The theory of the R matrix was developed in nuclear physics. As usually presented, the theory makes use of Green's theorem to relate value and slope of the radial channel orbitals at r_0, expanding these functions for $r < r_0$ as linear combinations of basis functions satisfying fixed boundary conditions at r_0. The true logarithmic derivative (or reciprocal of the R matrix in multichannel formalism) is computed from Green's theorem, despite the use of basis functions whose logarithmic derivatives at r_0 have a fixed but arbitrary value. Because of the inherent discontinuity of the boundary derivative, this expansion tends to converge slowly, but this can be improved by an approximate method due to Buttle (1967).

In nuclear physics, the specifically nuclear interaction is of short range, so r_0 can be chosen such that full scattering information is given by the R matrix. The method has been extended to electron–atom scattering, where long-range potentials are important, by combining basis expansion of the scattering wave function within r_0 with explicit numerical solution of coupled differential equations outside r_0 (Burke *et al.*, 1971; Burke and Robb, 1972; Burke, 1973; Burke and Robb, 1975). This method has the advantage of allowing processing of algebraic equations containing matrix elements of nonlocal operators within r_0 while exploiting the simple asymptotic form of the close-coupling equations (without exchange) outside r_0. This requires r_0

to be large enough to allow approximation of the optical potential by a matrix local potential outside r_0.

The R matrix can be matched at r_0 to external channel orbitals, assumed to be known exactly, to determine the K matrix. With the coefficient matrices α_i in the canonical form indicated by Eq. (1.16), the exact radial channel orbital in channel p is given by Eq. (1.5) as

$$f_{ps}(r) = \sum_q [w_{0pq}(r)\delta_{qs} + w_{1pq}(r)K_{qs}], \tag{2.118}$$

where the asymptotic forms of functions in channel p are

$$w_{0pq}(r) \sim k_p^{-1/2} \sin(k_p r - \tfrac{1}{2}l_p\pi)\delta_{pq},$$
$$w_{1pq}(r) \sim k_p^{-1/2} \cos(k_p r - \tfrac{1}{2}l_p\pi)\delta_{pq}. \tag{2.119}$$

The arguments here must be modified for Coulomb or dipole scattering. These functions are defined by inward integration of the coupled equations valid for $r > r_0$, with asymptotic boundary conditions as indicated. Equations (2.117) and (2.118) can be solved for the K matrix

$$K_{st} = -\sum_p \left[w_{1ps}(r_0) - \sum_q R_{pq}r_0 w'_{1qs}(r_0) \right]^{-1}_{sp}$$
$$\times \left[w_{0pt}(r_0) - \sum_q R_{pq}r_0 w'_{0qt}(r_0) \right]_{pt}, \tag{2.120}$$

in terms of the R matrix at r_0. Equations (2.119) and (2.120) as written refer to open channels only. When external closed-channel orbitals are included in the region outside r_0, the R matrix is indexed by both open and closed channels, and a single function that vanishes as $r \to \infty$ must be included in Eq. (2.119) for each closed channel. In Eq. (2.120), the indices p, q then refer to all channels, but the indices s, t refer to open channels only. The first factor in the matrix product on the right is the s, p rectangular submatrix of the inverse of a matrix obtained by extending both indices to include closed channels.

The theory of the R-matrix method can be understood most clearly in a variational formulation. In the single-channel case, the essential derivation was given by Kohn (1948) as a variational principle for the logarithmic derivative of the radial channel orbital at r_0. If h is the radial Hamiltonian operator, the variational functional is

$$\Xi = \int_0^\infty f(h - \varepsilon)f \, dr. \tag{2.121}$$

If $f(r)$ is an exact wave function for $r > r_0$, this reduces to the functional considered by Kohn (1948),

$$\Xi_\lambda = \int_0^{r_0} f(h - \varepsilon)f\,dr$$

$$= \int_0^{r_0} \left[\tfrac{1}{2}(f')^2 + (V - \tfrac{1}{2}k^2)f^2\right] dr - \tfrac{1}{2}\lambda f^2(r_0), \tag{2.122}$$

obtained by integrating the kinetic energy term by parts. The parameter λ is the logarithmic derivative

$$\lambda = f'(r_0)/f(r_0) = 1/Rr_0, \tag{2.123}$$

defining R as the single-channel R matrix.

For an infinitesimal variation δf, the variation of Ξ_λ, treating λ as a fixed parameter, is

$$\delta\Xi_\lambda = 2\int_0^{r_0} \delta f(h - \varepsilon)f\,dr + \delta f(r_0)[f'(r_0) - \lambda f(r_0)]. \tag{2.124}$$

This vanishes for unconstrained variations of f if and only if

$$(h - \varepsilon)f = 0, \qquad 0 \le r \le r_0, \tag{2.125}$$

and

$$\lambda = f'(r_0)/f(r_0). \tag{2.126}$$

If $f(r)$ is approximated by a finite expansion in linearly independent basis functions $\{\eta_a\}$,

$$f = \sum_{a=1}^{n} \eta_a c_a, \tag{2.127}$$

then variation of Ξ_λ with respect to the coefficients c_a gives the matrix equation

$$\sum_b A_{ab}c_b = \tfrac{1}{2}\lambda\eta_a(r_0)f(r_0), \tag{2.128}$$

where

$$A_{ab} = \int_0^{r_0} \left[\tfrac{1}{2}\eta_a'\eta_b' + \eta_a(V - \tfrac{1}{2}k^2)\eta_b\right] dr. \tag{2.129}$$

Kohn (1948) pointed out that λ is uniquely determined as a condition for the existence of a solution of Eq. (2.128), but he did not put this solution in the usual form of the R-matrix theory. This can easily be done by writing Eq. (2.128) as

$$c_a = \tfrac{1}{2}\lambda \sum_b (A^{-1})_{ab}\eta_b(r_0)f(r_0). \tag{2.130}$$

Hence

$$f(r_0) = \sum_a \eta_a(r_0)c_a = \tfrac{1}{2}\lambda \sum_a \sum_b \eta_a(r_0)(A^{-1})_{ab}\eta_b(r_0)f(r_0), \qquad (2.131)$$

which requires that

$$R = (r_0\lambda)^{-1} = \frac{1}{2r_0}\sum_a \sum_b \eta_a(r_0)(A^{-1})_{ab}\eta_b(r_0), \qquad (2.132)$$

the standard formula for the single-channel R matrix.

This early derivation by Kohn has an important feature that was not pointed out at the time but has recently been recognized to be of great practical significance. Kohn's derivation places no condition on the basis functions $\{\eta_a\}$ other than linear independence. The nonvariational derivation, which makes use of Green's theorem, requires the introduction of a complete set of basis functions in the interval $0 \le r \le r_0$. Such sets are obtained as eigenfunctions of some model Hamiltonian, with a fixed boundary condition

$$r_0\eta_a'(r_0) = b\eta_a(r_0) \qquad (2.133)$$

at r_0 (Wigner and Eisenbud, 1947; Lane and Thomas, 1958; Burke and Robb, 1975). In general, the value λ of the logarithmic derivative computed from Eq. (2.132) will differ from the imposed value b/r_0. If $f(r)$, for $r \le r_0$, is expanded as a linear combination of functions η_a with a fixed boundary condition as given by Eq. (2.134), it itself must satisfy the same boundary condition. This means that the function $f(r)$ in the standard R-matrix theory has a discontinuity of slope at r_0. This discontinuity leads to slow convergence of the basis set expansion and necessitates use of the Buttle correction (Buttle, 1967). An alternative method (Fano and Lee, 1973; Lee, 1974) is to adjust the parameter b so that it agrees with λ, which requires an iterative calculation.

The variational derivation of Eq. (2.132), as given here, requires only that the basis functions η_a should be linearly independent and vanish at $r = 0$. The variational expression, Eq. (2.124), shows that *both* conditions Eqs. (2.125) and (2.126) must be satisfied for Ξ_λ to be stationary. This shows that a "complete" set of functions $\{\eta_a\}$ with fixed boundary condition at r_0 corresponding to $b \ne \lambda$ is not in fact complete for the representation of $f(r)$. Such a basis set is complete for the interior interval $0 \le r < r_0$ but not at the boundary $r = r_0$. Either this basis set must be augmented, as in Buttle's method, to include a function not satisfying the fixed boundary condition, or functions must be used that are complete over a larger interval $0 \le r < r_0 + \Delta r_0$ for a fixed boundary condition at the upper limit.

Equation (2.132) is valid even if a fixed boundary condition is imposed. This is because of the form of A_{ab}, Eq. (2.129), which can be expressed as the matrix representation of an operator

$$(h - \varepsilon) + \tfrac{1}{2}\delta(r - r_0)(d/dr). \tag{2.134}$$

The second term here, the Bloch operator (Bloch, 1957), both symmetrizes A_{ab} and removes the dependence on any specific boundary condition. Direct use of the functional Ξ_λ is equivalent to introducing the Bloch operator. An interesting consequence is that Eq. (2.132), which appears from the usual derivation to be valid only for the special case of $b = 0$, is in fact the general expression for R, independent of any imposed boundary condition.

Kohn's basic derivation, which leads to Eq. (2.132) and appears in the same paper (Kohn, 1948) as his well-known variational principle given here as Eq. (2.56), was not recognized in some of the subsequent literature on R-matrix theory. Several alternative forms of the variational theory have been successively proposed or rediscovered.

In another early paper, Jackson (1951) derives a multichannel variational principle for the R matrix. Jackson's method can be applied to the generalized close-coupling equations, Eq. (1.38), with the assumption that exact solutions, given for $r \geq r_0$, are normalized by specifying their values at r_0 in terms of the R matrix. If the two linearly independent solutions in open channel p are denoted by $u_{0ps}(r)$ and $u_{1ps}(r)$, respectively, for a global close-coupling solution indexed by s, the assumed boundary conditions at r_0 are

$$
\begin{aligned}
u_{0ps}(r_0) &= 0, & u'_{0ps}(r_0) &= \delta_{ps}r_0^{-1/2}, \\
u_{1ps}(r_0) &= \delta_{ps}r_0^{+1/2}, & u'_{1ps}(r_0) &= 0.
\end{aligned}
\tag{2.135}
$$

Equation (2.118) is replaced by

$$f_{ps}(r) = \sum_t (u_{0pt}\delta_{ts} + u_{1pt}R_{ts}), \qquad r \geq r_0. \tag{2.136}$$

No distinction is made between open and closed channels, but the eventual transformation to obtain the K matrix requires setting the coefficients of exponentially increasing terms in the closed-channel functions to zero. From Eqs. (2.135) and (2.136),

$$
\begin{aligned}
f_{ps}(r_0) &= R_{ps}r_0^{1/2}, \\
f'_{ps}(r_0) &= \delta_{ps}r_0^{-1/2},
\end{aligned}
\tag{2.137}
$$

so that Eq. (2.117) is satisfied.

Since exact solutions of the close-coupling equations are assumed outside r_0, Eq. (2.23) becomes

$$\Xi_{st} = \sum_p \sum_q \int_0^{r_0} f_{ps}(r)m^{pq}f_{qt}(r)\, dr, \tag{2.138}$$

and the variation of Ξ is given in matrix notation by Eq. (2.25). The present boundary conditions, Eqs. (2.135) and (2.136), give for the final term in Eq. (2.25), after integration by parts and using Eqs. (2.137),

$$-\tfrac{1}{2}\sum_p [f_{ps}(r_0)\delta f'_{pt}(r_0) - \delta f_{pt}(r_0)f'_{ps}(r_0)] = \tfrac{1}{2}\delta R_{st}. \qquad (2.139)$$

As in the case of Kohn's variational principle, Eqs. (2.26) and (2.139) imply that the Jackson functional

$$[R] = R - 2\Xi^\dagger \qquad (2.140)$$

is stationary with respect to variations about an exact solution of the close-coupling equations. Moreover, if $[R]$ is stationary with respect to all variations, then the wave function is exact. For linear variations of the wave function, it will be shown that the stationary value of Ξ is zero. Then Eq. (2.140) shows that the computed R matrix is stationary.

The multichannel generalization of Eq. (2.122), in matrix notation, is

$$\Xi = f^\dagger A f - \frac{1}{2r_0}f^\dagger(r_0)R^{-1}f(r_0), \qquad (2.141)$$

where A includes a diagonal Bloch operator, as in Eq. (2.134). If expanded in a linearly independent set of basis functions $\eta_a(r)$, such that

$$f_{ps}(r) = \sum_{a=1}^n \eta_a(r)c_a^{ps}, \qquad (2.142)$$

the functional becomes

$$\Xi_{st}^\dagger = \sum_a \sum_b \sum_p \sum_q c_a^{pt}\left[A_{ab}^{pq} - \frac{1}{2r_0}\eta_a(r_0)(R^{-1})_{pq}\eta_b(r_0)\right]c_b^{qs}, \qquad (2.143)$$

if all functions and coefficients are real. Variation of $\Xi_{st}^\dagger$ with respect to the coefficients c_a^{pt}, for fixed R, equivalent to varying $[R]$ as defined by Eq. (2.140), gives the linear equations

$$\sum_q \sum_b A_{ab}^{pq}c_b^{qs} = \frac{1}{2r_0}\eta_a(r_0)\sum_q (R^{-1})_{pq}\sum_b \eta_b(r_0)c_b^{qs} \qquad (2.144)$$

$$= \frac{1}{2r_0}\eta_a(r_0)\sum_q (R^{-1})_{pq}f_{qs}(r_0). \qquad (2.145)$$

The formal solution is

$$c_a^{ps} = \frac{1}{2r_0}\sum_t \sum_b (A^{-1})_{ab}^{pt}\eta_b(r_0)\sum_q (R^{-1})_{tq}f_{qs}(r_0) \qquad (2.146)$$

or

$$f_{ps}(r_0) = \sum_a \eta_a(r_0)c_a^{ps}$$

$$= \sum_t \frac{1}{2r_0} \sum_a \sum_b \eta_a(r_0)(A^{-1})_{ab}^{pt}\eta_b(r_0) \sum_q (R^{-1})_{tq}f_{qs}(r_0). \qquad (2.147)$$

This implies that

$$R_{pq} = \frac{1}{2r_0} \sum_a \sum_b \eta_a(r_0)(A^{-1})_{ab}^{pq}\eta_b(r_0). \qquad (2.148)$$

Since the use of the Bloch operator in A_{ab}^{pq} replaces kinetic energy terms by the symmetrized form $\frac{1}{2}\eta'_a\eta'_b$, as in Eq. (2.129), the matrices A and R are both symmetric. By a slight generalization of the present derivation, different basis functions and channel radii can be used in each different channel. This would replace $\eta_a(r_0)$ in Eq. (2.148) by $\eta_{pa}(r_{0p})$, and similarly for $\eta_b(r_0)$.

It can easily be verified that Ξ vanishes if the coefficients c_a^{ps} are determined by Eq. (2.146). It follows from Eq. (2.140) that the computed R_{pq} is identical with the stationary functional $[R]$.

Jackson (1951) points out that the assumed variational trial function, expanded in a "complete" basis set with fixed boundary condition at r_0, in general has a discontinuity in its first derivative at r_0. Although Jackson states that this discontinuity is unavoidable, in fact the derivation requires only linear independence of the basis functions and is valid for basis functions with arbitrary boundary values and slope at r_0. Jackson's derivation, which provides the full multichannel generalization of Kohn's variational principle, Eq. (2.124) for the single-channel logarithmic derivative, has been largely ignored in subsequent literature.

Bloch (1957) considered several theories of nuclear reactions, including the Wigner–Eisenbud R-matrix theory. Although Bloch did not consider variational derivations of the R matrix, he introduced the boundary derivative operator, indicated here in Eq. (2.134), to define an Hermitian effective Hamiltonian operator in the Green's function theory. In discussing basis set expansions, Bloch considered only complete sets defined by fixed boundary conditions at r_0.

Lane and Robson (1969), apparently unaware of Jackson's work, introduced a symmetrized variational expression that differs from that used by Kohn (1948) and Jackson (1951). This symmetrized variational functional was found not to be stationary when applied to trial functions expressed as linear combinations of fixed basis functions. Although this work was the first to mention the use of unrestricted basis functions, the apparent failure of the stationary property probably discouraged later attempts to use such functions.

Oberoi and Nesbet (1973a) explicitly introduced the use of basis functions not restricted by a fixed boundary condition at r_0 as a device for improving convergence. The multichannel Kohn variational principle was modified so that trial functions with discontinuous derivatives at r could be used. For basis functions without fixed logarithmic derivatives at r_0, the resulting formula for the R matrix requires inversion of an unsymmetric matrix, but the computed R matrix is itself symmetric. When a fixed boundary condition is imposed on the basis functions, the variational expression was shown to reduce to the usual formula. Model calculations showed rapid convergence without use of the Buttle correction when unconstrained simple basis functions were used.

This variational derivation of the R matrix for arbitrary basis functions was shown by Oberoi and Nesbet (1974) to be equivalent to an alternative method developed independently by Schlessinger and Payne (1974), using the Bloch operator, so that all matrices are symmetric. In this form, the variational theory is equivalent to that of Jackson (1951), except that the external channel functions are defined by their asymptotic forms rather than by boundary conditions at r_0. The variational derivation (Oberoi and Nesbet, 1973a, 1974) shows that the K matrix computed in R-matrix theory is stationary with respect to variations of the expansion coefficients c_a^{ps}.

Fano and Lee (1973) used a variational functional equivalent to Eq. (2.122) or (2.141) in a novel way. The variational principle was used at fixed energy to derive the eigenvalues of the R matrix and a corresponding set of eigenvectors. By iteration, the logarithmic derivatives at r_0 of basis functions used to compute R can be adjusted to agree with the calculated eigenvalues of $(Rr_0)^{-1}$. This makes it possible to remove the discontinuities in radial derivatives at r_0.

In the method for electron–atom scattering described by Burke and Robb (1975), n basis functions $\eta_{0,a}$ with a specified fixed boundary condition at r_0 are obtained as eigenfunctions of a model Hamiltonian H_0. The matrix H_{ab} is diagonalized as $E_a\delta_{ab}$. The eigenvectors define functions η_a, orthonormal over $0 \le r \le r_0$, as linear combinations of the original basis functions. An additional function $\phi_0(r, E)$ is obtained by integrating the Schrödinger equation for H_0 outward to r_0 at the given value of E. Then, using single-channel scattering as an example, the R matrix for H_0 is

$$R_0(E) = \left[\frac{r_0\phi_0'(r_0, E)}{\phi_0(r_0, E)} - b\right]^{-1}, \tag{2.149}$$

where b is the fixed logarithmic derivative of the basis orbitals at r_0. When Eq. (2.149) is combined with the computed R matrix, given by Eq. (2.132) but with the matrix H_{ab} diagonalized, the R matrix as corrected by Buttle (1967) is

$$R(E) = \frac{1}{2r_0} \sum_{a=1}^{n} \frac{\eta_a^2(r_0)}{E_a - E} + R_0(E) - \frac{1}{2r_0} \sum_{a=1}^{n} \frac{\eta_{0,a}^2(r_0)}{E_{0,a} - E}. \tag{2.150}$$

Here, following the standard Green's function theory, the R matrix is defined by

$$f(r_0) = R[r_0 f'(r_0) - bf(r_0)], \tag{2.151}$$

which differs from Eq. (2.117) unless $b = 0$.

When the Buttle correction is included, the approximate function $f(r)$ is a linear combination of the n basis functions $\{\eta_{0,a}\}$ and the additional function $\phi_0(r, E)$. Zvijac *et al.* (1975) pointed out that a further improvement could be obtained by determining the final linear combination of these functions variationally. Their variational correction is that given by the Kohn formula, using a Bloch operator in the Hamiltonian as in Eq. (2.134). In the general variational formalism described here, their final result could be obtained more directly by adding $\phi_0(r, E)$ to the original basis set before evaluating Eq. (2.132).

Shimamura (1977) has reviewed model calculations with various alternative R-matrix methods. Calculations using basis functions without fixed boundary conditions show greatly improved convergence compared to the traditional method. The need for generalized basis functions has been recognized in the nuclear physics literature (Purcell, 1969). Chatwin and Purcell (1971) derive a generalization of the variational principle for the logarithmic derivative in one dimension (Kohn, 1948) and use variational basis functions that are not orthogonal over $0 \leq r \leq r_0$ in model calculations. The use of such basis functions is discussed in the context of the finite-element method by Nordholm and Bacskay (1978).

2.6. Hybrid Methods

The R-matrix method has been implemented for electron–atom scattering in the close-coupling formalism, with polarization pseudostates (Burke and Robb, 1975). In this context, the method takes advantage of the simple asymptotic form of the long-range potentials in the close-coupling equations. There are two essential difficulties that may limit the range of applicability of this close-coupling approach. The first is that for calculations of electron scattering by complex atoms, especially for inelastic scattering, the number of coupled channels required may be beyond the practical capabilities of the method. The second problem is that the R-matrix method cannot deal with nonlocal potentials that extend outside the channel radius r_0. This problem becomes especially serious in considering excitation of Rydberg states since these target atom states have effective radii proportional to the square of the principal quantum number. Nonlocal exchange potentials cannot be neglected unless r_0 is outside the target state electronic distribution.

In order to avoid these limitations, it would be desirable to have a method that can exploit the simple form of coupled ordinary differential equations in the asymptotic region, while avoiding a complete change of representation at an artificial channel boundary.

Methods that combine numerical integration of coupled differential equations for large r with an algebraic or matrix variational procedure can be called *hybrid* methods. The R-matrix method is a method of this kind. Another approach, proposed by Oberoi and Nesbet (1973b), uses the formalism of the matrix variational method but replaces the continuum basis functions, whose asymptotic forms are given by Eqs. (2.1), with functions whose asymptotic forms are defined by solutions of a model scattering problem. This method of numerical asymptotic functions (NAF) will be described in detail here.

Another general hybrid method can be based on the variational principle of Schwinger (1947), which is described in detail in standard references (in particular, Moiseiwitsch, 1966, pp. 256–261). This method can be used to compute the K matrix implied by a formally exact solution of the generalized close-coupling equations, expressed in terms of the Green's function defined by a model scattering problem. The formalism required for applications to multichannel scattering will be derived in Section 2.7.

In the numerical asymptotic function method (Oberoi and Nesbet, 1973b), exact solutions are assumed to be known for the system of differential equations, in matrix notation,

$$(h - \varepsilon)w = 0, \tag{2.152}$$

which define a model close-coupling problem. The generalized matrix radial operator m^{pq} defined by Eq. (1.15) is then expressed in the form

$$m = h - \varepsilon + \Delta V, \tag{2.153}$$

where the multichannel difference potential ΔV^{pq} is in general a nonlocal operator. If the model operator h^{pq} contains only local potentials, analytic solutions of Eq. (2.152) may be known, or the coupled differential equations may be solved by numerical integration. Exact solutions are required only for $r \geq r_0$, whose r_0 is some channel radius. It is not necessary to impose orthogonality constraints on the model channel functions w, so Eq. (2.152) can be considered without the inhomogeneous Lagrange multiplier terms required to enforce such constraints. These exact model channel orbitals are matched at r_0 to interior functions. Two linearly independent interior functions, both regular at $r = 0$ but otherwise arbitrary, are required for every radial channel in order to match both value and slope of an exact model channel orbital at r_0. This construction defines continuum basis functions that replace the functions $F_{ip}(r)$ of Eqs. (2.1). These functions are orthogonalized to all quadratically integrable basis orbitals included in a particular calculation. The orthogonalized functions are used in an appropriately modified variational calculation of the K matrix.

For a multichannel model Hamiltonian, the continuum basis functions of Eqs. (2.1) must be generalized to a matrix form. As in Eqs. (2.119), the radial functions are defined by inward integration, with asymptotic forms in open channel p

$$w_{0pq}(r) \sim k_p^{-1/2} \sin\left[\theta_p(r)\right]\delta_{pq},$$
$$w_{1pq}(r) \sim k_p^{-1/2} \cos\left[\theta_p(r)\right]\delta_{pq},$$

(2.154)

where, to include the case of electron–ion scattering,

$$\theta_p(r) = k_p r - \tfrac{1}{2} l_p \pi + \frac{Z}{k_p} \ln\left(2k_p r\right) + \arg \Gamma\left(l_p + 1 - \frac{i}{k_p} Z\right). \quad (2.155)$$

Here Ze is the ionic charge and Γ is the gamma function. For a closed channel, k_p becomes $i\kappa_p$ and the radial asymptotic form for inward integration is

$$w_{Cpq}(r) \sim \kappa_p^{-1/2} \exp\left(-\kappa_p r\right)\delta_{pq}. \quad (2.156)$$

Only a single asymptotic function is needed for a closed channel since the exponentially increasing component must be eliminated.

Exact regular solutions of the model close-coupling equations define a model K matrix as the real symmetric matrix

$$(\tan \Delta)_{pq} = \sum_\sigma x_{p\sigma} x_{q\sigma} \tan \eta_\sigma \quad (2.157)$$

expressed in eigenchannel form. Expressed in terms of the matrices

$$(\cos \Delta)_{pq} = \sum_\sigma x_{p\sigma} x_{q\sigma} \cos \eta_\sigma \quad (2.158)$$

and

$$(\sin \Delta)_{pq} = \sum_\sigma x_{p\sigma} x_{q\sigma} \sin \eta_\sigma, \quad (2.159)$$

open-channel components of exact model wave functions outside the channel radius r_0 are of the form

$$f_{ps}(r) = \sum_q \left[w_{0pq}(r)(\cos \Delta)_{qs} + w_{1pq}(r)(\sin \Delta)_{qs}\right]. \quad (2.160)$$

The Kohn variational method can be used to compute an effective K matrix in the basis of the model open-channel functions if Eq. (2.21) is valid for the open-channel matrix m_{ij}^{pq} computed in that basis. This requires an appropriate definition of solutions of the model close-coupling equations that would be irregular if carried into $r = 0$. Such solutions are

$$g_{ps}(r) = \sum_q [-w_{0pq}(r)(\sin \Delta)_{qs} + w_{1pq}(r)(\cos \Delta)_{qs}] \qquad (2.161)$$

if the regular solutions are those of Eq. (2.160). These "irregular" solutions are actually made regular at the origin in the NAF method by matching onto regular functions at r_0. The open-channel part of the asymptotic coefficient matrix for channel orbitals defined by Eqs. (2.160) and (2.161) is

$$u = \begin{bmatrix} \cos \Delta & -\sin \Delta \\ \sin \Delta & \cos \Delta \end{bmatrix}, \qquad (2.162)$$

defined as a transformation from the diagonal channel representation appropriate to Eqs. (2.154).

Equation (2.21) is obviously valid for the diagonal channel representation since the asymptotic surface integrals in matrices m_{01} and m_{10} are the same as those for noninteracting scattering channels. The matrix u of Eq. (2.162) is an orthogonal matrix of the canonical form given by Eq. (2.85) for transformations that preserve the antisymmetric component of m_{ij}^{pq}, as expressed by Eq. (2.21). It follows immediately that the functions defined by Eqs. (2.160) and (2.161), matched onto arbitrary regular interior functions at r_0, can be used as open-channel basis functions for matrices m' in a modified Kohn variational formula. This gives a stationary matrix $[K']$, from Eq. (2.111), in the model channel function basis. The true K matrix is given by Eq. (2.115), where C and S are the matrices defined by Eqs. (2.158) and (2.159), respectively.

When the model close-coupling system includes closed channels, the open-channel matrix m_{ij}^{pq} can be generalized to

$$\begin{bmatrix} (h' - \varepsilon)_{OO} & (h' - \varepsilon)_{OC} \\ (h' - \varepsilon)_{CO} & (h' - \varepsilon)_{CC} \end{bmatrix}, \qquad (2.163)$$

where the subscripts of O and C refer to open and closed channel components, respectively. The coefficient matrix u is extended to include closed-channel coefficients α_C, β_C and an additional column vector γ, orthogonal to α and β, for each closed channel. Explicit solution of the linear equations for the closed-channel coefficients α_C, β_C gives

$$u_{CO} = (\alpha, \beta)_C = -(h' - \varepsilon)_{CC}^{-1}(h' - \varepsilon)_{CO}. \qquad (2.164)$$

When substituted back into the linear equations for the open-channel coefficients, this results in equations of the usual form, Eq. (2.17), for open channels only. The effective open-channel matrix m_{ij}^{pq} is given by the partitioned matrix expression

$$m = (h' - \varepsilon)_{OO} - (h' - \varepsilon)_{OC}(h' - \varepsilon)_{CC}^{-1}(h' - \varepsilon)_{CO}. \qquad (2.165)$$

In the multichannel NAF method matrix elements of $h' - \varepsilon$ are to be computed by the appropriate generalization of Eq. (2.9). The generalization consists of modifying Eqs. (2.11) and (2.12) so that the bound–free matrix becomes

$$M_{\mu,is} = \left(\Phi_\mu | H - E | \sum_p \mathscr{A}\Theta_p\phi_{ips} \right) \qquad (2.166)$$

and the free–free matrix becomes

$$M_{ij}^{st} = \left(\sum_p \mathscr{A}\Theta_p\phi_{ips} | H - E | \sum_q \mathscr{A}\Theta_q\phi_{jqt} \right). \qquad (2.167)$$

The open-channel orbital functions ϕ_{0ps} and ϕ_{1ps} are defined by their radial factors, $r^{-1}f_{ps}$ and $r^{-1}g_{ps}$, respectively, as given by Eqs. (2.160) and (2.161). Orbital functions ϕ_{Cps} for closed channels are defined by radial factors expressed as linear combinations of $r^{-1}w_{Cpq}(r)$, from Eq. (2.156), with coefficients corresponding to the vectors γ in the extended transformation matrix u. The modified Eq. (2.9) gives the matrix $h' - \varepsilon$ as in Eq. (2.163), including both open- and closed-channel submatrices. Then the reduced open-channel matrix m is computed from Eq. (2.165) and used in the modified Kohn formula for $[K']$, from which the K matrix is computed, as given by Eq. (2.115).

Analysis given in Section 2.4 shows that the Kohn variational formalism for $[K']$ in the transformed basis is valid in any representation of the model open-channel functions that can be obtained from Eq. (2.162) by a transformation of the canonical form indicated in Eq. (2.85). By appropriate choice of the model close-coupling Hamiltonian, $[K']$ may be reduced in magnitude since $(\tan \Delta)$ in Eq. (2.157) is an approximation to the true K matrix. The diagonal representation of Eq. (2.154) may also be used, so that $(\tan \Delta)$ is a unit matrix, and $[K']$ becomes identical with $[K]$. However, this may not take full advantage of the model approximation to $[K]$ unless a variational method is used that considers a general preliminary transformation of m_{ij}^{pq}.

If the model equation, Eq. (2.152), solved for $r \geq r_0$, is used to simplify the integrands of the bound–free and free–free integrals, Eqs. (2.11) and (2.12), only the difference potential ΔV of Eq. (2.153) remains in the radial integrand outside r_0. If the channel radius is chosen as in the R-matrix method so that the asymptotic potentials are accurate outside r_0, the radial integrals over continuum orbitals reduce to integrals for $r \leq r_0$ only. By eliminating the long-range oscillatory part of the integrand, this method can greatly simplify the numerical evaluation of these integrals even if the difference potential is not completely negligible outside r_0.

An alternative representation of the solutions of the model close-coupling equations is obtained by multiplying Eq. (2.162) on the right by the matrix

$$\begin{bmatrix} X & 0 \\ 0 & X \end{bmatrix}, \tag{2.168}$$

where X is the orthogonal matrix of eigenchannel column vectors defined by Eq. (2.157). The resulting representation replaces Eqs. (2.160) and (2.161), respectively, by

$$f_{p\sigma}(r) = \sum_q [w_{0pq}(r)x_{q\sigma} \cos \eta_\sigma + w_{1pq}(r)x_{q\sigma} \sin \eta_\sigma],$$

$$\tag{2.169}$$

$$g_{p\sigma}(r) = \sum_q [-w_{0pq}(r)x_{q\sigma} \sin \eta_\sigma + w_{1pq}(r)x_{q\sigma} \cos \eta_\sigma].$$

From Eqs. (2.154), the asymptotic forms, characteristic of eigenchannels, are

$$f_{p\sigma}(r) \sim k_p^{-1/2} x_{p\sigma} \sin [\theta_p(r) + \eta_\sigma],$$

$$g_{p\sigma}(r) \sim k_p^{-1/2} x_{p\sigma} \cos [\theta_p(r) + \eta_\sigma]. \tag{2.170}$$

Oberoi and Nesbet (1973b) tested the NAF method by single-channel calculations of e^-–H scattering in the static exchange approximation and of e^-–He$^+$ scattering in the static approximation (which omits exchange integrals). The latter calculation, with a Coulomb scattering potential, avoids the two-electron integrals over Coulomb wave functions that have been one of the major difficulties preventing extension of algebraic variational methods to electron–ion scattering. In both sets of calculations phase shifts calculated for several values of k were in excellent agreement with exact values. Different channel radii and angular momenta and two different representations of channel orbitals inside r_0 were considered. It was found that the simple monomials r^{l+1} and r^{l+2} could be used as basis functions inside r_0 but gave less satisfactory convergence, especially for Coulomb scattering at relatively large k values (up to $1.895a_0^{-1}$), than did oscillatory functions of the form

$$(1 - e^{-r})^{l+1} \sin kr,$$

$$(1 - e^{-r})^{l+1} \cos kr. \tag{2.171}$$

The NAF method has not yet been applied to multichannel scattering. The choice of representation of channel functions inside r_0 is quite arbitrary. One possibility would be to use the regular solutions of the model equations for all r, modifying only the irregular solutions inside r_0.

Rountree and Parnell (1977) have proposed a hybrid method that uses the local potential part of the close-coupling equations as a model potential. Only the regular model solution is used. The regular free-particle function is taken to be the other linearly independent continuum basis function. Quite accurate s-wave singlet phase shifts were computed for e^-–H scattering in the static exchange approximation. Because the "irregular" solutions of the model equations are not used, the chosen continuum basis cannot represent accurate solutions for long-range potentials unless the incremental phase shift due to the nonlocal interactions is small. For this reason, it may not be possible, in general, to avoid the use of the "irregular" solutions for large r.

2.7. The Schwinger Variational Principle

The variational principle of Schwinger (1947) makes use of the Lippmann–Schwinger (1950) integral equation, equivalent to the Schrödinger equation, to derive a stationary expression for the scattering amplitude. The formalism can also be expressed in terms of the K matrix, or $\tan \eta$ for single-channel scattering (e.g., Moiseiwitsch, 1966, pp. 256–261; Joachain, 1975, pp. 233–238).

The single-channel radial Schrödinger equation, in terms of a model Hamiltonian operator h, is

$$(h - \varepsilon + \Delta V)f(r) = 0 \tag{2.172}$$

where

$$h = -\frac{1}{2}\frac{d^2}{dr^2} + \frac{l(l+1)}{2r^2} + V(r), \tag{2.173}$$

$$\varepsilon = \tfrac{1}{2}k^2, \tag{2.174}$$

and $\Delta V(r)$ is a difference potential as in Eq. (2.153). It is assumed that regular and irregular solutions $w_0(r)$ and $w_1(r)$ are known such that

$$(h - \varepsilon)w_i(r) = 0, \tag{2.175}$$

with asymptotic boundary conditions as in Eqs. (2.154). These functions can be used to construct the *principal value Green's function* G, a linear integral operator whose kernel is

$$g(r, r') = \begin{cases} 2w_0(r)w_1(r'), & r < r', \\ 2w_1(r)w_0(r'), & r > r'. \end{cases} \tag{2.176}$$

Acting on any function $F(r)$ that is regular at the origin $(\sim r^{l+1})$ and vanishes at infinity, G has the property

$$(h - \varepsilon)GF = F, \tag{2.177}$$

where

$$GF(r) = 2\left[w_0(r) \int_r^\infty w_1(r') + w_1(r) \int_0^r w_0(r') \right] F(r')\, dr'. \tag{2.178}$$

Equation (2.177) can be verified by applying the operator $(h - \varepsilon)$ to Eq. (2.178), making use of the Wronskian relation

$$w_1(r)w_0'(r) - w_1'(r)w_0(r) = \text{const} = 1, \tag{2.179}$$

which follows from Eq. (2.175) and is evaluated from the asymptotic forms of w_0 and w_1. The kernel function $g(r, r')$ is symmetric in r, r', regular at the origin in either variable, and continuous. It has discontinuous first derivative in either variable at $r = r'$, which gives $(h - \varepsilon)G$ the properties of a Dirac δ function. The asymptotic form of GF is

$$GF \sim 2w_1(r) \int_0^\infty w_0(r')F(r')\, dr', \qquad r \to \infty, \tag{2.180}$$

while for small r

$$GF \sim 2w_0(r) \int_0^\infty w_1(r')F(r')\, dr', \qquad r \to 0. \tag{2.181}$$

Thus GF is regular at the origin but is asymptotically proportional to the irregular function $w_1(r)$.

Equation (2.172) has a formal solution in terms of the model Green's function,

$$f = w_0 - G\Delta Vf. \tag{2.182}$$

This integral equation for f is the Lippmann–Schwinger (1950) equation. It is equivalent to the Schrödinger equation plus the boundary condition for large r,

$$f(r) \sim w_0 + w_1 \tan \eta, \tag{2.183}$$

where, from Eq. (2.180),

$$\tan \eta = -2 \int_0^\infty w_0(r')\Delta V(r')f(r')\, dr. \tag{2.184}$$

It is assumed that ΔVf is regular at the origin and that ΔV vanishes at infinity more rapidly than r^{-2}. Coulomb or static dipole potentials, asymptotically proportional to r^{-1} and r^{-2}, respectively, should be included in the model

Hamiltonian h. Equation (2.184) gives a formula for $\tan \eta$ in terms of the exact wave function $f(r)$. The asymptotic boundary condition built into the Lippmann–Schwinger equation can be modified by using a different form of the irregular function in the definition of $g(r, r')$, Eq. (2.176). The Wronskian relation, Eq. (2.179), is unaltered if any constant multiple of w_0 is added to w_1. The particular choice made here, for the principal value Green's function, gives $\tan \eta$ directly or the K matrix when generalized to multichannel scattering. The phase shift given by Eq. (2.183) is defined relative to the phase shift of the model problem and should be denoted by η' if $V(r)$ in Eq. (2.173) is a model potential. The multichannel generalization gives K', the effective K matrix in the model channel function basis, from which the K matrix is computed as in Eq. (2.115).

Equation (2.182) can be considered to give w_0 as a linear functional of f,

$$w_0 = (I + G\Delta V)f, \tag{2.185}$$

where I is the identify operator. If this is substituted into Eq. (2.184) it gives the bilinear functional

$$\tan \eta = -2(f|\Delta V + \Delta VG\Delta V|f) \tag{2.186}$$

in an obvious notation. This also is an exact expression for $\tan \eta$ if f is an exact wave function with the correct asymptotic boundary condition. In the variational method of Schwinger (1947), Eqs. (2.184) and (2.186) are combined to give a homogeneous formula

$$[\tan \eta] = -2(w_0|\Delta V|f)(f|\Delta V + \Delta VG\Delta V|f)^{-1}(f|\Delta V|W_0), \tag{2.187}$$

that is stationary with respect to variations of $f(r)$ about an exact wave function.

The Schwinger formula, Eq. (2.187), has several remarkable properties. Because $[\tan \eta]$ is homogeneous, f can be multiplied by an arbitrary constant. Moreover, f occurs in the formula only in the form ΔVf, which vanishes at infinity because of the properties of ΔV. This implies that there is no constraint on the asymptotic form of variational trial functions. The Schwinger variational principle in fact determines f only in the range of the potential ΔV. If this range is defined by a channel radius r_0, then Eq. (2.182), assumed in the Schwinger theory, determines $f(r)$ outside this range as

$$f(r) = w_0(r) - 2w_1(r) \int_0^{r_0} w_0(r')\Delta V(r')f(r')\, dr'$$

$$= w_0(r) + w_1(r) \tan \eta, \qquad r \geq r_0. \tag{2.188}$$

It can be seen that the Schwinger variational principle, used with a model Hamiltonian such that ΔV can be neglected outside some r_0, provides a particularly elegant hybrid theory. A variational approximation to $f(r)$ is required only for $r \leq r_0$, and only regular functions occur in the theory.

The stationary property of the Schwinger formula follows immediately from Eqs. (2.184), (2.186), and (2.187). For an infinitesimal variation of f about an exact wave function,

$$\delta[\tan \eta] = \left[\frac{(\delta f|\Delta V|w_0)}{(f|\Delta V|w_0)} - \frac{(\delta f|\Delta V + \Delta VG\Delta V|f)}{(f|\Delta V + \Delta VG\Delta V|f)} + \text{transpose} \right][\tan \eta]$$

$$= -2(\delta f \Delta V|w_0 - f - G\Delta Vf) + \text{transpose}. \tag{2.189}$$

This implies that Eq. (2.182), the Lippmann–Schwinger equation, is valid in the range of ΔV if and only if $[\tan \eta]$ is stationary. Equation (2.182) determines $f(r)$ outside this range, as in Eq. (2.188).

If $f(r)$ is expanded in a basis of linearly independent functions $\{\eta_a\}$ as in Eq. (2.127), ΔV can be expanded in the same basis as a linear integral operator with the kernel

$$\Delta v(r, r') = \sum_a \sum_b \eta_a(r) \Delta V_{ab} \eta_b(r'), \tag{2.190}$$

where

$$f(r) = \sum_a \eta_a(r)(a|f), \tag{2.191}$$

defining the coefficients $(a|f)$. For $[\tan \eta]$ to be stationary with respect to variation of these coefficients, Eq. (2.189) implies

$$(a|w_0 - f - G\Delta Vf) = 0, \qquad \text{all } a. \tag{2.192}$$

Then

$$(a|f) = \sum_b (I + G\Delta V)_{ab}^{-1}(b|w_0) \tag{2.193}$$

$$= \sum_b \sum_c (\Delta V + \Delta VG\Delta V)_{ab}^{-1} \Delta V_{bc}(c|w_0). \tag{2.194}$$

If Eq. (2.191), with these values of the coefficients $(a|f)$, is substituted into either Eq. (2.184), (2.186), or (2.187) they each give the same result,

$$[\tan \eta] = -2 \sum_a \sum_b \sum_c \sum_d (w_0|a)\Delta V_{ac}(\Delta V + \Delta VG\Delta V)_{cd}^{-1} \Delta V_{db}(b|w_0). \tag{2.195}$$

If the kernel of ΔV could be represented exactly by Eq. (2.190), this derivation shows that the coefficients $(a|f)$ are given exactly by Eq. (2.194). The part of f orthogonal to the basis $\{\eta_a\}$ is also determined exactly by Eq. (2.182), but it is not relevant to $[\tan \eta]$ because of the assumed completeness of the expansion of ΔV. This argument has been used by Rescigno *et al.*

(1974a, b; 1975) to derive a hybrid method, the *T-matrix method*, that has been applied to electron–molecule scattering (Fliflet and McKoy, 1978). Because this method solves the scattering problem defined by the truncated ΔV exactly, no spurious singularities should occur in Eqs. (2.194) and (2.195).

As originally presented, the T-matrix method solved the Lippmann–Schwinger equation in a matrix representation, using the Green's function appropriate to the transition matrix, T_{pq} of Eq. (1.24), or $e^{i\eta} \sin \eta$ for a single channel. The T matrix is evaluated from a formula analogous to Eq. (2.184). The free-particle Green's function was used, corresponding to assuming $V(r) = 0$ in Eq. (2.173). The present derivation extends this method by considering a general model Hamiltonian h, which could contain all long-range potentials in a particular application, and by establishing the relationship to the Schwinger variational principle. This shows that $[\tan \eta]$ as computed by Eq. (2.195) is stationary in the same sense as the Kohn functional. Fliflet and McKoy (1978) have reformulated the original method in terms of the K matrix, as is done here. Inaccuracy due to truncating the representation of the scattering potential is variationally corrected by using the Kohn functional for $[\tan \eta]$ with the full wave function computed from Eq. (2.182).

A hybrid variational method could make use of GF as defined by Eq. (2.178) for some arbitrary function $F(r)$, together with the regular model channel orbital $w_0(r)$, as continuum basis functions, using the Hulthén–Kohn variational principle. If $F(r)$ vanished outside some channel radius r_0 and was normalized so that

$$2 \int_0^{r_0} w_0(r')F(r') \, dr' = 1, \tag{2.196}$$

this method would be a special case of the NAF method described in Section 2.6 above. The multichannel formalism given there could be used directly.

Certain choices of $F(r)$ give especially simple results. If $F(r)$ has a simple analytic form, $GF(r)$ might be obtained by analytic integration of Eq. (2.177) as a differential equation. For example, if $F(r)$ is λr^{l+1}, vanishing outside r_0, with

$$\lambda = \left[2 \int_0^{r_0} w_0(r')(r')^{l+1} \, dr' \right]^{-1}, \tag{2.197}$$

then $GF(r)$ is the particular solution of Eq. (2.177) for which

$$\frac{[GF(r_0)]'}{GF(r_0)} = \frac{w_1'(r_0)}{w_1(r_0)}. \tag{2.198}$$

If $F(r)$ is taken to be proportional to ΔVf, which is then expanded in the basis $\{\eta_a\}$, the method becomes equivalent to the Schwinger method.

For practical use in electron–atom scattering theory, the Schwinger variational principle must be generalized to a multichannel formalism. In order to do this, several special problems must be solved. The necessary formalism will be derived here. A Green's function will be defined that provides a multichannel version of the Lippmann–Schwinger equation. The orthogonality conditions used to abstract a one-electron multichannel theory (the generalized close-coupling equations) from the basic $(N + 1)$-electron problem must be taken into account in applying this formalism. This is done here, first, by defining the concept of *transfer invariance*, a property of the $(N + 1)$-electron scattering wave function, and then by showing that the generalized close-coupling equations have the related property of *transfer covariance* and are equivalent to a transfer covariant Lippmann–Schwinger equation. The resulting Schwinger variational functional will be expressed in terms of matrix elements of $(N + 1)$-electron operators so that bound-state computational techniques can be used.

The system of equations

$$\sum_q m^{pq} u_{qs}(r) = 0, \qquad (2.199)$$

where the radial operator m^{pq} is defined by Eq. (1.15), can be solved in terms of the Green's function of a model problem

$$\sum_q (h - \varepsilon)^{pq} w_{iq\sigma}(r) = 0, \qquad i = 0, 1, \qquad (2.200)$$

where

$$m^{pq} = (h - \varepsilon)^{pq} + \Delta V^{pq} \qquad (2.201)$$

defining a difference potential or linear operator ΔV. To simplify the analysis, it will be assumed that closed-channel components of the model problem have been eliminated by a partitioning transformation, as in Eqs. (2.164) and (2.165) and in the close-coupling equations. Then $(h - \varepsilon)^{pq}$, in general, contains a nonlocal potential, but the solutions are known exactly. The channel indices in Eq. (2.201) then refer to open channels only. The model functions $w_{ip\sigma}$ are defined to correspond to eigenchannel asymptotic boundary conditions, as in Eqs. (2.169). Here $w_{0p\sigma}$ is the regular solution $f_{p\sigma}$, and $w_{1p\sigma}$ is the solution $g_{p\sigma}$, irregular at $r = 0$.

The multichannel model Green's function appropriate to the eigenchannel representation is the linear operator G^σ_{pq}, defined by its kernel

$$g^\sigma_{pq}(r, r') = 2w_{0p\sigma}(r)w_{1q\sigma}(r'), \qquad r < r',$$

$$= 2w_{1p\sigma}(r)w_{0q\sigma}(r'), \qquad r > r'. \qquad (2.202)$$

If $\{F_q(r)\}$ is a vector of functions, quadratically integrable and regular at the origin, defined in each open channel q, then

$$\sum_\sigma \sum_q G^\sigma_{pq} F_q(r) = \sum_\sigma \sum_q 2 \left\{ w_{0p\sigma}(r) \int_r^\infty w_{1q\sigma}(r') + w_{1p\sigma}(r) \int_0^r w_{0q\sigma}(r') \right\} F_q(r') \, dr',$$

$$(2.203)$$

such that

$$\sum_{p'} \sum_q \sum_\sigma (h - \varepsilon)^{pp'} G^\sigma_{p'q} F_q(r) = F_p(r). \tag{2.204}$$

The proof of this result follows that of Eq. (2.177), except that the Wronskian relationship is generalized to

$$\sum_p [w'_{ip\sigma}(r) w_{jp\tau}(r) - w_{ip\sigma}(r) w'_{jp\tau}(r)] = \delta_{\sigma\tau} (\delta_{i0}\sigma_{j1} - \delta_{i1}\delta_{j0}), \tag{2.205}$$

which follows from Eqs. (2.169) and (2.170).

The model Green's function gives solutions of Eq. (2.199) in the form of the Lippmann–Schwinger equation

$$u_{p\tau} = w_{0p\tau} - \sum_{p',q,\sigma} G^\sigma_{pp'} \Delta V^{p'q} u_{q\tau}. \tag{2.206}$$

From Eq. (2.202), the asymptotic form of the Green's function term here is a linear combination of functions $w_{1p\sigma}$, with coefficients

$$K'_{\sigma\tau} = -2 \sum_{p,q} (w_{0p\sigma} | \Delta V^{pq} | u_{q\tau}). \tag{2.207}$$

This is the effective K matrix in the model eigenchannel representation. The true K matrix is computed as in Eq. (2.115), with $C_{p\sigma}$ and $S_{p\sigma}$ given, respectively, by $x_{p\sigma} \cos \eta_\sigma$ and $x_{p\sigma} \sin \eta_\sigma$, from Eqs. (2.169).

To derive a Schwinger variational functional, Eq. (2.206) can be written, in a matrix notation, as

$$w_0 = u + G\Delta V u. \tag{2.208}$$

This is substituted into Eq. (2.207) to give

$$K' = -2(u|\Delta V + \Delta VG\Delta V|u). \tag{2.209}$$

The multichannel Schwinger functional is given by the *matrix* product

$$[K'] = -2(w_0|\Delta V|u)(u|\Delta V + \Delta VG\Delta V|u)^{-1}(u|\Delta V|w_0). \tag{2.210}$$

This is stationary for variations of u if and only if u is an exact solution of Eq. (2.199). The proof follows exactly as in Eq. (2.189), if the order of matrix products is maintained in the derivation. If u_{ps} is expanded as

$$u_{ps}(r) = \sum_a \eta_{pa}(r)(pa|ps), \tag{2.211}$$

where $\{\eta_{pa}\}$ are radial basis functions for orbital angular momentum l_p, then $[K']$ is a function of the coefficients $(pa|ps)$. As in Eq. (2.195), the stationary expression obtained by varying $[K']$ with respect to these coefficients is

$$[K']_{\sigma\tau} = -2 \sum_{a,b} \sum_{p,p',q,q'} (w_{0p\sigma}|\Delta V|\eta_{p'a})$$

$$\times [(\eta_{p'a}|\Delta V + \Delta VG\Delta V|\eta_{q'b})]^{-1}_{p'a,q'b}(\eta_{q'b}|\Delta V|w_{0q\tau}). \quad (2.212)$$

Equation (2.199) differs from the generalized close-coupling equations, given by Eq. (1.38), because $u_{ps}(r)$ is not required to be orthogonal to the basis functions $\{\eta_{pa}(r)\}$ and because the Lagrange multiplier term is absent. In order to consider the relationship between these equations, it is necessary to return to the original representation of the $(N+1)$-electron wave function Ψ_s, given by Eq. (1.1). A consistency condition required in constructing Ψ_s is that the $(N+1)$-electron basis $\{\Phi_\mu\}$ must be complete for representation of any function of the form $\mathscr{A}\Theta_p\phi_{pa}$, where Θ_p is any target state function included in the first term $\sum_p \mathscr{A}\Theta_p\psi_{ps}$ of Ψ_s, and ϕ_{pa} is an orbital function whose radial factor $r^{-1}\eta_{pa}(r)$ is formed from any η_{pa} in the radial basis set for orbital angular momentum l_p. This consistency condition ensures that Ψ_s has a property that can be called *transfer invariance*. This means that Ψ_s is not altered by adding any multiple of ϕ_{pa} to the channel orbital ψ_{ps}, so long as a suitable adjustment is made to the coefficients $c_{\mu s}$. This process simply transfers a quadratically integrable term $\mathscr{A}\Theta_p\phi_{pa}$ between the two sums $\sum_p \mathscr{A}\Theta_p\psi_{ps}$ and $\sum_\mu \Phi_\mu c_{\mu s}$. In deriving the generalized close-coupling equations, transfer invariance is exploited to define a particular canonical form, in which the radial channel orbitals $f_{ps}(r)$ are orthogonal to all basis functions $\{\eta_{pa}(r)\}$ in the set used to construct $\{\Phi_\mu\}$. This orthogonality condition provides a unique definition of the functions $f_{ps}(r)$.

Transfer invariance of Ψ_s implies that a valid system of close-coupling equations exists for any arbitrary allocation of terms $\mathscr{A}\Theta_p\phi_{pa}$ between the two sums in Eq. (1.1). For any such allocation, there will be a specific component of ϕ_{pa} in the channel orbital ψ_{ps} and a specific set of coefficients $\{c_{\mu s}\}$. This consistency, which can be called *transfer covariance*, is ensured by solving Eq. (1.36) for the coefficients $c_{\mu s}$, which become linear functionals of the channel orbitals ψ_{ps}, as do the optical potential terms in m^{pq}, indicated explicitly as the last term of Eq. (1.41). The result of this analysis is that the generalized close-coupling equations are transfer covariant if Eq. (1.36) is used to determine the matrix optical potential in m^{pq}.

Because Ψ_s is transfer invariant, the coefficients $c_{\mu s}$ are not uniquely determined. Any multiple of ϕ_{pa} added to ψ_{ps} adds the same multiple of

$$-(\Phi_\mu|H - E|\mathscr{A}\Theta_p\phi_{pa}) \quad (2.213)$$

to the right-hand side of Eq. (1.36) and the equivalent linear combination of matrix elements $(\Phi_\mu|H - E|\Phi_\nu)$ to the left-hand side. Requiring ψ_{ps} to be orthogonal to all ϕ_{pa} enforces uniqueness but introduces Lagrange multipliers into the close-coupling equations. An alternative is to require Eq. (1.37) to be valid and to adjust the Lagrange multipliers accordingly. Equation (1.37) is compatible with

$$(\mathscr{A}\Theta_p\phi_{pa}|H - E|\Psi_s) = 0, \qquad \text{all } p, a, \qquad (2.214)$$

which reduces to

$$\sum_q (\eta_{pa}|m^{pq}|f_{qs}) = 0, \qquad \text{all } p, a, \qquad (2.215)$$

if Eqs. (1.35) and (1.36) are used to determine m^{pq} through Eq. (1.15). Equations (2.214) and (1.35) are compatible since $\mathscr{A}\Theta_p\phi_{pa}$ is a linear combination of the functions $\{\Phi_\mu\}$. From Eq. (1.38), however, Eq. (2.215) is just the condition that all Lagrange multipliers should vanish, so the close-coupling equations reduce to Eq. (2.199) in this case.

This argument shows that Eq. (2.199), without Lagrange multipliers, is a member of the class of transfer covariant equations equivalent to the generalized close-coupling equations. Hence $[K']$ as given by Eq. (2.210) or (2.212) is a stationary estimate of the K matrix in the model eigenchannel representation and is a transfer invariant quantity. This argument also shows that the generalized close-coupling equations can be solved without Lagrange multipliers or orthogonality constraints so long as transfer covariance is maintained through the use of Eq. (1.36) with a transfer invariant total wave function.

The Lippmann–Schwinger equation, Eq. (2.206), which is equivalent to the modified close-coupling equation, Eq. (2.199), can be expressed in a transfer covariant form. The functions u_{ps} can be replaced by the orthogonalized functions f_{ps} of the original close-coupling equations if the difference potential ΔV is replaced by a transfer covariant linear operator $\Delta \tilde{V}$ that incorporates the Lagrange multipliers of Eq. (1.38), the generalized close-coupling equations. The resulting Lippmann–Schwinger equation is

$$f_{p\tau} = w_{0p\tau} - \sum_{p',q,\sigma} G^\sigma_{pp'} \Delta \tilde{V}^{p'q} f_{q\tau}. \qquad (2.216)$$

The effect of the Lagrange multipliers here is to adjust $\Delta \tilde{V}$ so that f is orthogonal to the basis set $\{\eta_{pa}\}$. Hence Eq. (2.216) is equivalent to the equation in matrix notation,

$$f = Pw_0 - PG\Delta\tilde{V}f, \qquad (2.217)$$

where P is the orthogonalizing projection operator. In terms of these definitions, the stationary Schwinger functional is the matrix product

$$[K'] = -2(\tilde{w}_0|\Delta\hat{V}|f)(f|\Delta\hat{V} + \Delta\hat{V}G\Delta\hat{V}|f)^{-1}(f|\Delta\hat{V}|w_0). \quad (2.218)$$

Because of transfer invariance, this must be identical with $[K']$ as given by Eq. (2.210). It should be noted that the *covariant operators* ΔV and $\Delta\hat{V}$ differ in the two cases.

In order to take advantage of the computational technique developed in N-electron bound-state theory, it is desirable to reformulate the Schwinger variational principle in terms of matrix elements of the full Hamiltonian H between $(N + 1)$-electron functions such as $\mathcal{A}\Theta_p\psi_{ps}$ or Φ_μ. The model operator $(h - \varepsilon)^{pq}$ of Eq. (2.200) is equivalent to an $(N + 1)$-electron operator $H_0 - E$ defined by its matrix elements

$$(\Phi_\mu|H_0 - E|\Phi_\nu) = \sum_p \sum_q (\Phi_\mu|\Theta_p Y_p)(h - \varepsilon)^{pq}(Y_q\Theta_q|\Phi_\nu), \quad (2.219)$$

where Y_p is a generalized orbital angular momentum and spin function that projects onto radial functions in channel p and includes the appropriate coupling with Θ_p to give specified total quantum numbers $LS\pi$. The one-electron multichannel difference potential ΔV^{pq} is replaced by $H - H_0$ if closed channels are included explicitly in $(h - \varepsilon)^{pq}$ and H_0. In order to derive a Lippmann–Schwinger equation, the multichannel Green's function must be generalized to include closed channels.

In defining the Green's function, two independent vector functions are required for each eigenchannel. For open channels, the functions $w_{0p\sigma}$ are regular at the origin, while the functions $w_{1p\sigma}$ are bounded at large r and defined so that the Wronskian is given by Eq. (2.205) as $\delta_{\sigma\tau}$. The same conditions can be applied for closed channels, starting from asymptotic forms in the diagonal channel representation of solutions of the model equations,

$$w_{0\gamma q}(r) \sim (2\kappa_\gamma)^{-1/2} \exp(\kappa_\gamma r)\delta_{\gamma q}, \quad (2.220)$$

$$w_{1\gamma q}(r) \sim (2\kappa_\gamma)^{-1/2} \exp(-\kappa_\gamma r)\delta_{\gamma q}. \quad (2.221)$$

These functions satisfy the Wronskian condition but in general are irregular at the origin.

If there are n_O open channels and n total channels, the model equations have n regular solutions $w_{0p\sigma}(r)$. Only n_O of them are bounded at infinity, and these will be denoted by open-channel values of the index σ. The $n - n_O$ remaining regular but unbounded solutions, which can be defined asymptotically by the first of Eqs. (2.220), will be denoted by closed-channel values of σ. Linearly independent irregular functions $w_{1p\sigma}(r)$ can be constructed so that they are all bounded at large r and so that the multichannel Wronskian is given by Eq. (2.205). This gives $2n$ linearly independent functions

$$w_{ip\sigma}(r) = \sum_{j=0}^{1} \sum_{q=1}^{n} w_{jpq}(r)u_{ji}^{q\sigma}, \qquad i = 0, 1; \quad p, \sigma = 1, \ldots, n.$$

$$(2.222)$$

For an eigenchannel representation, the open-channel submatrix must have the form defined by Eqs. (2.169). A canonical form for the closed-channel coefficients is given by

$$u_{0i}^{q\sigma} = \delta_{q\sigma}\delta_{i0}, \qquad q > n_O, \tag{2.223}$$

and

$$u_{j1}^{q\sigma} = \delta_{q\sigma}\delta_{j1}, \qquad \sigma > n_O. \tag{2.224}$$

Equation (2.223) expresses the condition that functions unbounded at large r are excluded from $w_{ip\sigma}$, $i = 0, 1$, if $\sigma \le n_O$, and from $w_{1p\sigma}$ for $\sigma > n_O$. Equation (2.224) excludes terms that are oscillatory at large r from $w_{1p\sigma}$ for $\sigma > n_O$, so that these functions vanish at large r and do not affect the K matrix.

In the diagonal channel representation, the Wronskian given by the model equations is

$$\sum_s (w'_{isp}w_{jsq} - w_{isp}w'_{jsq}) = \delta_{pq}(\delta_{i0}\delta_{j1} - \delta_{i1}\delta_{j0}) \tag{2.225}$$

$$= J_{ij}^{pq}, \tag{2.226}$$

where J is the antisymmetric matrix of Eq. (2.84), extended to include closed channels. The coefficient matrix $u_{ji}^{q\sigma}$ is chosen to satisfy the Wronskian condition in the eigenchannel representation,

$$\sum_s (w'_{is\sigma}w_{js\tau} - w_{is\sigma}w'_{js\tau}) = \delta_{\sigma\tau}(\delta_{i0}\delta_{j1} - \delta_{i1}\delta_{j0}) \tag{2.227}$$

$$= J_{ij}^{\sigma\tau}. \tag{2.228}$$

This requires the matrix condition

$$u^\dagger J u = J. \tag{2.229}$$

The kernel of the Green's function G_{pq}^σ can be written in terms of the functions defined above as

$$g_{pq}^\sigma(r, r') = 2w_{0p\sigma}(r)w_{1q\sigma}(r'), \qquad r < r', \tag{2.230}$$

$$= 2w_{1p\sigma}(r)w_{0q\sigma}(r'), \qquad r > r', \tag{2.231}$$

where all indices refer to both open and closed channels. It can be verified that Eq. (2.204) holds for G_{pq}^{σ}, by the Ansatz

$$GF_p(r) = \sum_i \sum_{\sigma} w_{ip\sigma}(r)X_{i\sigma}(r), \tag{2.232}$$

with the auxiliary condition, required to give unique solutions,

$$\sum_i \sum_{\sigma} w_{ip\sigma}(r)X'_{i\sigma}(r) \equiv 0. \tag{2.233}$$

Substitution of GF_p as given by Eq. (2.232) into Eq. (2.204), together with Eq. (2.233), gives matrix equations for $X'_{i\sigma}(r)$ that can be solved explicitly using the Wronskian relationship of Eq. (2.228). The integrated form of this result, incorporating appropriate boundary conditions at small and large r, is given by Eq. (2.203).

Equation (1.37), from which the generalized close-coupling equations are derived, can be written in terms of the $(N + 1)$-electron model Hamiltonian H_0 of Eq. (2.219) as

$$(\Theta_p|H_0 - E|\Psi_{\sigma}) = -(\Theta_p|H - H_0|\Psi_{\sigma}). \tag{2.234}$$

A formal solution is given by the Lippmann–Schwinger equation

$$\Psi_{\sigma} = \sum_p \mathscr{A}\Theta_p Y_p \left[w_{0p\sigma} - \sum_{\tau,q} G_{pq}^{\tau}(Y_q\Theta_q|H - H_0|\Psi_{\sigma}) \right], \tag{2.235}$$

using notation as in Eq. (2.219), in terms of the Green's function G_{pq}^{σ} defined by Eq. (2.230). The first term in Eq. (2.235) satisfies the homogeneous equation

$$\left(Y_p\Theta_p|H_0 - E|\sum_q \mathscr{A}\Theta_q Y_q w_{0q\sigma} \right) = \sum_q (h - \varepsilon)^{pq} w_{0q\sigma} = 0, \tag{2.236}$$

from Eq. (2.219), the definition of H_0. From the defining property of the Green's function, Eq. (2.204), the second term in Eq. (2.235) is a particular solution of Eq. (2.234). This verifies that Eq. (2.235) implies Eq. (1.37) and hence is equivalent to the generalized close-coupling equations in nonorthogonalized form, if the coefficients $\{c_{\mu\sigma}\}$ are determined by Eq. (1.36) as usual.

A transfer invariant radial channel orbital function $u_{p\sigma}(r)$ can be defined by

$$u_{p\sigma}(r) = (Y_p\Theta_p|\Psi_{\sigma}). \tag{2.237}$$

From Eq. (2.235) the asymptotic behavior of this function defines, for open channels σ and τ,

$$K'_{\sigma\tau} = -2\sum_p (\mathscr{A}\Theta_p Y_p w_{0p\sigma}|H - H_0|\Psi_{\tau}) \tag{2.238}$$

as the K matrix in the model eigenchannel representation. Here H_0 is assumed to be symmetrized over electron coordinates so the antisymmetrizing operator $\mathscr{A}$ can be used as with H. Matrix elements are defined as in Eq. (2.219). Substituting for the term in w_0 from Eq. (2.235), the corresponding bilinear formula is

$$K'_{\sigma\tau} = -2\Bigg[(\Psi_\sigma | H - H_0 | \Psi_\tau)$$
$$+ \sum_{p,q} (\Psi_\sigma | H - H_0 | \Theta_p Y_p) \sum_p G^\sigma_{pq} (Y_q \Theta_q | H - H_0 | \Psi_\tau) \Bigg], \qquad (2.239)$$

and the stationary Schwinger functional is

$$[K']_{\sigma\tau} = -2 \sum_{\sigma'\tau'} \Delta V_{\sigma\sigma'} (\Delta V + \Delta V G \Delta V)^{-1}_{\sigma'\tau'} \Delta V_{\tau'\tau} \qquad (2.240)$$

in an obvious matrix notation, where ΔV now denotes $H - H_0$ and the Green's function is

$$G = \sum_{p,\sigma,q} |\Theta_p Y_p > G^\sigma_{pq} < Y_q \Theta_q| \qquad (2.241)$$

in Dirac notation.

As usual in the Schwinger formalism, the variational principle only determines the wave function in the range of ΔV since the residual wave function is determined by the Lippmann–Schwinger equation. If this formalism is used as a hybrid theory, the model Hamiltonian can be chosen to include all asymptotic long-range potentials. Hence it is valid to expand Ψ_σ solely in the quadratically integrable basis $\{\Phi_\mu\}$ for use as a variational trial function in the Schwinger functional, and to determine the coefficients $c_{\mu\sigma}$ by the functional's stationary property. From Eq. (2.240), this leads by the usual argument to

$$[K']_{\sigma\tau} = -2 \sum_{\mu,\nu,p,q} (\mathscr{A}\Theta_p Y_p w_{0p\sigma} | H - H_0 | \Phi_\mu)$$
$$\times [(\Phi_\mu | H - H_0 + (H - H_0)G(H - H_0) | \Phi_\nu)]^{-1}_{\mu\nu}$$
$$\times (\Phi_\nu | H - H_0 | \mathscr{A}\Theta_q Y_q w_{0q\tau}), \qquad (2.242)$$

where G is defined by Eq. (2.241).

The matrix elements here can be simplified. Since $w_{0p\sigma}$ is an exact solution of the model equation, $H - H_0$ is equivalent to $H - E$ in the first and last matrix elements of Eq. (2.242), which reduce to bound-free integrals as in Eq. (2.11),

$$(\Phi_\mu | H - H_0 | \mathscr{A}\Theta_p \phi_{0p\sigma}) = (\Phi_\mu | H - E | \mathscr{A}\Theta_p \phi_{0p\sigma})$$
$$= M_{\mu,0p\sigma}, \qquad (2.243)$$

where $r^{-1} w_{0p\sigma}$ is the radial factor of orbital $\phi_{0p\sigma}$.

The intermediate matrix element in Eq. (2.242) is of the same form as that in Eq. (2.210),

$$(u|\Delta V + \Delta V G \Delta V|u),\qquad(2.244)$$

where

$$\Delta V = m - (h - \varepsilon)\qquad(2.245)$$

and $(h - \varepsilon)$ acts as if it were the inverse operator G^{-1}. Then Eq. (2.244) simplifies to

$$(u|m + mGm - (h - \varepsilon) + (h - \varepsilon)G(h - \varepsilon) - mG(h - \varepsilon) - (h - \varepsilon)Gm|u)$$
$$= (u|mGm - m|u).\qquad(2.246)$$

By the same argument, the matrix element in Eq. (2.242) reduces to

$$(\Phi_\mu|H - H_0 + (H - H_0)G(H - H_0)|\Phi_\nu)$$
$$= (\Phi_\mu|(H - E)G(H - E)|\Phi_\nu) - (\Phi_\mu|H - E|\Phi_\nu),\qquad(2.247)$$

where the second term is the bound–bound matrix element $M_{\mu\nu}$, as in Eq. (2.10).

The first term in Eq. (2.247) can be defined as

$$(MGM)_{\mu\nu} = \sum_{p,\sigma,q} (\Phi_\mu|H - E|\Theta_p Y_p)G^\sigma_{pq}(Y_q\Theta_q|H - E|\Phi_\nu).\qquad(2.248)$$

If radial functions are defined such that

$$v_{q\nu}(r) = (Y_q\Theta_q|H - E|\Phi_\nu)\qquad(2.249)$$

then Eq. (2.248) reduces to the double integral

$$(MGM)_{\mu\nu} = \sum_{p,\sigma,q} \int_0^\infty dr \int_0^\infty dr'\, v_{p\mu}(r)g^\sigma_{pq}(r, r')v_{q\nu}(r'),\qquad(2.250)$$

where $g^\sigma_{pq}(r, r')$ is defined by Eqs. (2.230) in terms of the model channel functions.

The final form of Eq. (2.242) is

$$[K']_{\sigma\tau} = -2\sum_{\mu,\nu}\sum_{p,q} M_{0p\sigma,\mu}[(MGM)_{\mu\nu} - M_{\mu\nu}]^{-1}_{\mu\nu}M_{\nu,0q\tau}.\qquad(2.251)$$

Computationally, this is similar in structure to m_{ij}^{pq}, as given by Eq. (2.9), which must be evaluated in the Kohn method. The Schwinger method, as formulated here, does not require free–free integrals or bound–free integrals for irregular continuum basis functions, but does require the evaluation of Green's function integrals $(MGM)_{\mu\nu}$, one for each element of the bound–bound matrix.

The fact that the one-electron Green's function is the formal inverse of the model operator can be used to express the kernel function in an alternative form (Fano, 1961) used in the single-channel case by Rescigno *et al.* (1974a,b; 1975),

$$g^{\sigma}_{pq}(r, r') = \frac{2}{\pi} P \int \frac{d\varepsilon'}{\varepsilon' - \varepsilon} w_{0p\sigma}(\varepsilon'; r) w_{0q\sigma}(\varepsilon'; r'). \qquad (2.252)$$

This follows from

$$\sum_{\sigma} \sum_{p'} (h - \varepsilon)^{pp'} g^{\sigma}_{p'q} = 2/\pi \sum_{\sigma} \int d\varepsilon' w_{0p\sigma}(\varepsilon'; r) w_{0q\sigma}(\varepsilon'; r')$$

$$= \delta(r, r')\delta_{pq}, \qquad (2.253)$$

using the completeness of the model wave functions and their asymptotic normalization. This form of the Green's function is established only for an open-channel model operator.

Although the Schwinger variational principle has been rarely used for electron–atom scattering calculations, the formalism as presented here offers several apparently viable alternatives to other methods. As a hybrid theory, when all important long-range multichannel potentials are included in the model Hamiltonian, the difference potential may be of sufficiently short range that expansion of the relevant part of the wave function in a quadratically integrable basis may converge efficiently. This could be exploited in either of two forms of the theory presented here: the multi-channel one-electron formalism, with a matrix optical potential, using Eq. (2.212), or the $(N + 1)$-electron formalism using Eq. (2.251).

The stationary multichannel Schwinger functionals given in Eqs. (2.210) and (2.240) may themselves be directly useful in providing a variational distorted-wave theory or in defining variational corrections to approximate close-coupling calculations.

3

Resonances and Threshold Effects

Introduction

Much of the energy-dependent structure observed in low-energy electron scattering arises from resonances and threshold effects. Resonances are due to nearly bound states of the $(N + 1)$-electron system of target atom and external electron. In a time-dependent formalism, such a state decays with a finite lifetime if it interacts with states in an electron scattering continuum. The uncertainty principle implies a corresponding energy width. Cross sections in the interacting continuum are modified within the energy range defined by the resonance energy and width.

At the energy threshold for excitation of each successive target atom state, the opening up of a new continuum affects the scattering in channels already open below threshold. This leads to energy-dependent scattering structures that have been observed in electron–atom scattering. Just above a threshold, scattered electrons of very low energy or wave number k are produced. Analytic formulas, valid for small values of k in a particular scattering channel, can be derived for scattering matrix elements and for the resulting cross sections. These formulas are useful in fitting experimental data and in explaining observed energy-dependent threshold structures.

In the present chapter, the theory of resonance and threshold effects will be presented, with emphasis on aspects relevant to electron scattering by neutral atoms. This theory is used in practice to relate computed K matrices to energy-dependent structures that can be compared with experimental data. The effects on these structures due to perturbation splitting of target atom energy levels will also be discussed, in terms of an energy modified adiabatic theory. Certain conclusions of this theory are applicable to excitation of rotational and vibrational states in electron–molecule scattering.

The general theory of scattering resonances and threshold effects is given by Newton (1966) and Joachain (1975). Aspects of the theory relevant to electron–atom scattering are discussed by Geltman (1969), Brandsden (1970), and Burke (1977). Reviews of theory and experimental data are given by Burke (1968), Taylor (1970), and Schulz (1973a,b). A detailed discussion of analytic properties of resonance and threshold effects, including many examples taken from model calculations, is given by McVoy (1967).

3.1. Electron Scattering Resonances

The simplest model of a scattering resonance is provided by a one-dimensional potential function $V(r)$ with an inner well enclosed by a potential barrier. If the value of $V(r)$ at large r is taken to be zero, the barrier height is some positive energy $\bar{E}$. The inner well may have true bound states, with $E_\alpha < 0$, but positive energy levels are also possible for $E_\alpha < \bar{E}$. If the barrier were of infinite width these would be true bound states, but in general the wave function must penetrate the barrier and match onto an oscillatory external function that extends to $r = \infty$. The internal part of such a wave function could be approximated by establishing an artificial boundary condition at some point within the barrier, for example, requiring functions to vanish at $\bar{r}$ corresponding to the barrier maximum $\bar{E}$. Eigenfunctions of the Schrödinger equation, subject to this boundary condition, corresponding to eigenvalues less than $\bar{E}$, would either be true bound states or approximations to positive energy states. In a time-dependent picture, the latter states would define wave packets whose time dependence could be described in terms of tunneling through the barrier. The amplitude within the barrier would decrease in time as the external oscillatory component built up. This model leads to exponential decay in time and is the basis of the quantum theory of spontaneous decay processes. The energy of a state decaying as $\exp(-t/\tau_\alpha)$ is imprecisely defined in the range $E_\alpha \pm \frac{1}{2}\Gamma_\alpha$, where the energy width Γ_α is $\hbar/\tau_\alpha$.

From the point of view of scattering theory, the phenomenon of interest is the modification of the external scattering continuum due to the influence of a decaying internal state. Except within the energy width Γ_α of the energy E_α of an internal function ϕ_α obtained with an artificial boundary condition as described above, the continuum wave functions can be taken to be orthogonal to ϕ_α. For energies E close to E_α, there is a strong perturbation, whose maximum effect occurs at some energy E_{res} (usually within the interval $E_\alpha \pm \frac{1}{2}\Gamma_\alpha$). The true wave function in this range is a linear combination of ϕ_α and continuum functions appropriate to the given energy E but with modified asymptotic phase. The detailed theory of this perturbation, given below, shows that for an isolated resonance, the continuum phase shift rises through π radians, increasing by $\pi/2$ radians in the interval $E_{res} \pm \frac{1}{2}\Gamma_\alpha$. This can result in a very prominent structural feature in the scattering cross section. The resonance energy E_{res} can be defined as the point of maximum rate of increase of the phase shift.

Again from the point of view of scattering theory, the physical process associated with a resonance is that of temporary capture into a positive energy internal state of the scattering potential or target system. In a time-dependent picture, an incoming wave packet is subject to a time delay related to the time constant τ_α.

For scattering in a single channel, the asymptotic radial wave function is given by Eq. (1.4). The time-dependent wave function has an additional factor $\exp(-i\omega t)$, where

$$\omega = \hbar k^2/2m = E/\hbar. \tag{3.1}$$

A wave packet constructed from stationary state wave functions has the general asymptotic form

$$u(r, t) \sim \int dk' a(k') \sin(k'r - \tfrac{1}{2}l\pi + \eta) e^{-i\omega t}, \tag{3.2}$$

where $a(k')$ can be assumed to define a peaked distribution centered on a fixed value of k corresponding to the incident energy $h^2k^2/2m$. Equation (3.2) can also be expressed in the form

$$u(r, t) \sim \int dk' a(k')\{\exp[-i(k'r + \omega t)] - (-1)^l \exp[i(k'r - \omega t + 2\eta)]\}, \tag{3.3}$$

as a superposition of time-dependent incoming and outgoing radial waves. At given time t, this integral is dominated by the value of k' corresponding to stationary phase of one of the two oscillatory terms. The effective motion of the wave packet can be followed by finding the value of $r(t)$ for which this value of k' coincides with the fixed value k that corresponds to the peak of the distribution function $a(k')$. If $t < 0$ the incoming wave has stationary phase if

$$r = -vt, \tag{3.4}$$

where

$$v = d\omega/dk = \hbar k/m. \tag{3.5}$$

This is the group velocity of the incident wave packet. If $t > 0$, the outgoing wave has stationary phase if

$$r = vt - 2\, d\eta/dk \tag{3.6}$$

$$= v(t - \Delta t), \tag{3.7}$$

defining a time delay

$$\Delta t = \frac{2m}{\hbar k}\frac{d\eta}{dk} \tag{3.8}$$

$$= 2\hbar \frac{d\eta}{dE}. \tag{3.9}$$

Equation (3.9) shows that energy-dependent structure in the phase shift and scattering cross section is directly related to a time delay in a time-dependent picture of the collision process. As originally shown by Wigner (1955), causality establishes a lower bound for Δt, which may be negative for repulsive potentials. For a potential of finite radius r_0, the minimum value of Δt allowed by causality is $-2r_0/v$ (Wigner, 1955). From Eq. (3.8), this implies

$$d\eta/dk \geq -r_0. \tag{3.10}$$

There is no upper limit on positive values of Δt. Thus the rate of decrease with energy of a phase shift is bounded, but there is no limit on the rate of increase, as required for a very narrow resonance.

The analytic theory of resonances, to be developed in more detail below, shows that for an isolated resonance in one channel the background phase shift is augmented by a resonance phase shift of the form

$$\Delta\eta = -\tan^{-1}(1/\varepsilon), \tag{3.11}$$

where

$$\varepsilon = (E - E_{\text{res}})/\tfrac{1}{2}\Gamma. \tag{3.12}$$

If the variation of the background phase shift is neglected near such a resonance, Eq. (3.9) gives

$$\Delta t = \frac{4\hbar}{\Gamma}\frac{d\eta}{d\varepsilon} = \frac{4\hbar}{\Gamma}\frac{1}{1+\varepsilon^2}. \tag{3.13}$$

If this is averaged over the resonance phase shift $\Delta\eta$, which increases from 0 to π over the full range of ε, the mean time delay is

$$\langle \Delta t \rangle = \frac{1}{\pi}\int_{-\infty}^{+\infty} \Delta t\, \frac{d\varepsilon}{1+\varepsilon^2} = \frac{2\hbar}{\Gamma}. \tag{3.14}$$

This is just twice the value of the time constant τ for the decaying internal state associated with the resonance.

For single-channel scattering, the S matrix of Eq. (1.23) reduces to $\exp(2i\eta)$. This factor occurs in Eq. (3.3) as the coefficient of the outgoing radial wave. At a resonance the factor in this coefficient due to the resonance is

$$\exp(2i\Delta\eta) = \frac{\varepsilon - i}{\varepsilon + i}, \tag{3.15}$$

from Eq. (3.11). If the scattering energy is analytically continued into the complex energy plane, Eq. (3.15) has a pole at

$$\varepsilon = -i \tag{3.16}$$

or

$$E = E_{\text{res}} - \tfrac{1}{2}i\Gamma. \tag{3.17}$$

The resonance energy E_{res} is the closest point on the real energy axis. Multichannel resonances are characterized in general by poles of the analytically continued S matrix at complex energies just below the positive real axis of E.

If E is real and negative, then k is replaced by $i\kappa$. The two terms in Eq. (3.3) contain factors $e^{+\kappa r}$ and $e^{-\kappa r}$, respectively. For bound states, the coefficient of the first term must vanish, or equivalently, the coefficient of the second term must be infinite. Thus the coefficient that corresponds to the S matrix continued to negative real energy values must have poles at bound-state energies. Both bound states and resonances are characterized by poles of the analytically continued S matrix, the latter at values in the lower half of the complex energy plane. The wave function corresponding to such a pole would depend on time through the factor

$$\exp[(E_{\text{res}} - \tfrac{1}{2}i\Gamma)t/i\hbar] = \exp(E_{\text{res}}t/i\hbar)\exp(-\Gamma t/2\hbar), \tag{3.18}$$

so that the probability density would decay in time according to the factor $\exp(-\Gamma t/\hbar)$. Hence the resonance width Γ and the lifetime of the associated decaying state are related by

$$\Gamma = \hbar/\tau. \tag{3.19}$$

Resonances in electron scattering by neutral atoms correspond to short-lived states of negative ions. The qualitative classification of such states is discussed by Taylor (1970). Resonances due to an effective potential with an inner well and external barrier are called *shape resonances*. Narrow resonances due to temporary capture of an electron by an excited state of the target atom are called *Feshbach resonances* (Feshbach, 1958,

1962). They occur at energies close to the excitation threshold of the parent target state. Taylor (1970) distinguishes two classes of Feshbach resonances. Those called core-excited type 1 (CE1) are below the relevant excitation threshold. They are very narrow because they cannot decay into the scattering continuum of the parent excited state. Core-excited resonances of type 2 (CE2) are above the parent state threshold but are bound temporarily by a centrifugal barrier. This requires $l > 0$ for the open scattering channels of the parent state. CE2 resonances can also be described as shape resonances, due to the centrifugal barrier, but the attractive part of the effective potential is the same as for CE1 resonances. Because decay into parent-state open channels is allowed, CE2 resonances may be broad compared with CE1 resonances, due essentially to the same binding mechanism. This may be direct Coulomb attraction to the ion core of the parent excited state or an effective static multipole or polarization potential due to the parent state.

An alternative classification of resonances can be made in terms of the nature of short-lived negative ion states (Nesbet, 1978a). The same classification is valid for bound or metastable states. An electron interacting with a neutral atom experiences a direct Coulomb attraction only if it penetrates inside the outer shell of target atom electrons. Hydrogenic orbital functions, appropriate to a Coulomb potential, have radial maxima that increase with the square of the principal quantum number. Hence screening of an ion core is nearly complete unless the external electron can occupy an orbital in an incomplete valence subshell. It is inconsistent to assign a principal quantum number with the usual interpretation to an external electron if it is outside the outermost valence shell since it does not experience a Coulomb potential.

When the external electron is incorporated into an incomplete valence subshell, this describes a valence state (ground or excited) of the negative ion. A resonance state of this structure can be called a *valence state* (VS) resonance (Nesbet, 1978a). These states are characterized by electronic configurations that have two or more electrons in the most loosely bound subshell. For example, the negative ions of open-shell atoms such as C, N, and O have ground configurations $1s^2 2s^2 2p^{d+1}$. The various LS-coupled states of these configurations occur either as true bound states, as scattering resonances, or as metastable states that cannot break up without violating $LS\pi$ symmetry. Although these resonances are usually categorized as shape resonances, the binding mechanism is not easily described by an effective one-electron potential function. The electrons in the outer $2p^{d+1}$ subshell are equivalent and are equally bound to the $1s^2 2s^2$ ion core, with incomplete screening.

Excited VS metastable states and resonances of rare gas negative ions are well known. In the case of He, configurations $1s2s^2$, $1s2s2p$, and $1s2p^2$ lead to such states in the energy range of the $1s2s$ and $1s2p$ excited states of the neutral atom. The narrow ^2S resonance at 19.36 eV, of mixed configuration $1s2s^2 + 1s2p^2$, corresponds to Taylor's classification CE1 since it lies below the $(1s2s)^3S$ threshold at 19.82 eV. The other resonances in this group would be classified as CE2 since they lie above the lowest parent-state threshold. Nevertheless, they are most directly analogous to the ground configuration "shape" resonances of open-shell negative ions. While the binding mechanism of incompletely screened attraction to an ion core is common to all VS states, the width or lifetime of VS resonances depends crucially on their location relative to parent-state thresholds. According to their location, VS states can be bound states, metastables, narrow resonances, or broad resonances. Taylor's classification is useful in distinguishing between the latter two possibilities.

Triply excited states $2s^2 2p$, $2s2p^2$, and $2p^3$ of He^- are also observed as resonances near the autoionizing neutral states of configurations $2s^2$, $2s2p$, or $2p^2$, which are approximately 35 eV above the ionization threshold of He. Thus, resonances with substantial delay times are observed for a system imbedded in an ionization continuum, which implies an infinite number of open two-electron channels. Corresponding triply excited states of doubly negative H have also been observed as resonances in $e^- -H^-$ scattering (Walton *et al.*, 1970; Taylor and Thomas, 1972).

If the external electron remains outside the outermost valence subshell of the target atom, it must be bound by an unusually strong multipole or polarization potential since the direct Coulomb potential is completely screened. The functional form of the radial wave function cannot be hydrogenic. Such states can be called nonvalence (NV) states. The best known example occurs for atomic hydrogen. The $2s$ and $2p$ excited levels are so nearly degenerate that they combine under the influence of an external electron to produce an effective long-range dipole potential (Gailitis and Damburg, 1963a). This leads to an infinite series of narrow resonances in $e^- -H$ scattering between 9.6 eV and the $n = 2$ threshold at 10.2 eV (Chen, 1967). In $e^- -He$ scattering, near the $1s3s$ and $1s3p$ thresholds, resonances occur that can be attributed to the temporary binding of an electron in the very strong electric dipole polarization potentials of these states (Nesbet, 1978a). These are examples of NV resonances.

Analytic continuation of the S matrix from positive to negative scattering energies is described by substituting $i\kappa$ for k. Physical bound states are described by S-matrix poles at positive values of κ. The opposite choice of sign, $-i\kappa$, describes an alternative analytic continuation of the S matrix onto a nonphysical image of the negative energy axis. Poles of the S matrix can occur on this nonphysical axis in the complex energy plane and must be considered in the same context as bound states or resonances. The states corresponding to such poles are called *virtual states*. Model calculations show that they occur for attractive potential wells without external barriers that are not quite strong enough to contain one additional bound level. As the attractive well depth is reduced, a bound state moves through the continuum threshold to become a virtual state if there is no external barrier, as in the case of s-wave scattering with a purely attractive potential. For $l > 0$, the centrifugal barrier causes the precursor bound state to become a true resonance as it passes through threshold. In electron scattering by He, virtual states have been identified, associated with the $(1s2s)^1S$ threshold (Burke *et al.*, 1969a) and also with the $(1s3s)^3S$ threshold (Nesbet, 1978a). The latter is identified as an NV virtual state in the attractive polarization potential of the $(1s3s)^3S$ parent state.

3.2. Theory of Resonances

For an isolated resonance in a single open channel, the phase shift is given by

$$\eta(E) = \eta_0(E) + \eta_{\text{res}}(E), \tag{3.20}$$

where

$$\tan \eta_{\text{res}} = -1/\varepsilon \tag{3.21}$$

with ε given by Eq. (3.12) in terms of the resonance energy E_{res} and width Γ. If η_0 is constant, Eq. (3.20) gives the resonant line shape formula (Fano, 1961; Fano and Cooper, 1965)

$$\sigma/\sigma_0 = (\varepsilon + q)^2/(1 + \varepsilon^2), \tag{3.22}$$

where

$$q = -\cot \eta_0. \tag{3.23}$$

Equation (3.22) gives a characteristically unsymmetrical resonance shape that goes to zero when $\varepsilon = -q$. In electron scattering, this partial cross section is superimposed on other partial wave contributions that in general would not show resonance structure at the same energy as E_{res}.

Equation (3.20) can be derived by application of the partitioning theory of Feshbach (1958, 1962), as in Eqs. (1.9)–(1.12). A simple example is provided by a shape resonance due to a one-dimensional potential function

$V(r)$, consisting of a potential well inside a finite barrier. The Hamiltonian is

$$H = -\frac{1}{2}\frac{d^2}{dr^2} + V(r). \tag{3.24}$$

Let ϕ_α be defined by an eigenvector of a finite matrix representation of H with eigenvalue E_α, using quadratically integrable basis functions. Then the scattering wave function can be expressed as

$$u(r) = f(r) + \phi_\alpha(r)c_\alpha. \tag{3.25}$$

Hence f is orthogonal to ϕ_α but otherwise unconstrained. In this application of the Feshbach theory, the Q space consists of the single function ϕ_α, while the P space is the orthogonal complement consisting of all functions orthogonal to ϕ_α. The Schrödinger equation in the Q space is

$$(\alpha|H - E|u) = 0, \tag{3.26}$$

or

$$c_\alpha = (E - E_\alpha)^{-1}(\alpha|H|f). \tag{3.27}$$

The Schrödinger equation in the P space is

$$(H - E)f - \phi_\alpha(\alpha|H - E|f) = -(H - E_\alpha)\phi_\alpha c_\alpha. \tag{3.28}$$

The background continuum function $\tilde{w}_0$ orthogonal to ϕ_α satisfies the integrodifferential equation

$$(H - E)\tilde{w}_0 - \phi_\alpha(\alpha|H - E|\tilde{w}_0) = 0. \tag{3.29}$$

An irregular solution $\tilde{w}_1$ of the same homogeneous equation can be defined, such that the Wronskian $\tilde{w}_1\tilde{w}_0' - \tilde{w}_0\tilde{w}_1'$ is unity. The asymptotic normalization is

$$\begin{aligned}
\tilde{w}_0 &\sim k^{-1/2}\sin(kr + \eta_0), \\
\tilde{w}_1 &\sim k^{-1/2}\cos(kr + \eta_0).
\end{aligned} \tag{3.30}$$

Here η_0 is the background phase shift that appears in Eq. (3.20). It should be noted that $\tilde{w}_0$ and $\tilde{w}_1$ are not the same functions as those obtained by orthogonalizing the continuum eigenfunctions w_0 and w_1 of $H - E$ to ϕ_α. Equation (3.29) must be solved by the method used for Eq. (1.38), as described in Eqs. (1.42) and (1.43).

Equation (3.28) can be solved in terms of a Green's function $\tilde{G}$, whose kernel $\tilde{g}(r, r')$ satisfies the inhomogeneous equation

$$(H - E)\tilde{g}(r, r') - \phi_\alpha(\alpha|H - E|\tilde{g}) = \delta(r, r') - \phi_\alpha(r)\phi_\alpha(r'), \tag{3.31}$$

subject to the condition that $(\alpha|\tilde{g})$ and $(\tilde{g}|\alpha)$ should vanish identically.

Equation (3.31) is the projection of the usual Green's function equation into the P space. Its solution can be obtained in terms of a modified Green's function $\bar{G}$ for the full operator $H - E$, with kernel

$$\bar{g}(r, r') = 2w_0(r_<)[w_1(r_>) + \mu w_0(r_>)], \tag{3.32}$$

where constants μ and λ are chosen so that

$$w_1 + \mu w_0 \sim \lambda \tilde{w}_1 \tag{3.33}$$

at large r. Here w_0 is the regular eigenfunction of $H - E$ with the usual normalization, and w_1 is the corresponding irregular function. Because Eq. (3.32) preserves the Wronskian normalization, $\bar{G}$ satisfies Eq. (2.177) but changes the asymptotic form of functions $\bar{G}F$. The kernel of $\hat{G}$ is then defined by

$$\tilde{g}(r, r') = \bar{g}(r, r') - \bar{G}\phi_\alpha(r)(\alpha|\bar{G}|\alpha)^{-1}\bar{G}\phi_\alpha(r'). \tag{3.34}$$

Equation (3.31) can be verified from the properties of $\bar{G}$. For any regular function $F(r)$, the asymptotic form of $\hat{G}F$ is $2\tilde{w}_1(\tilde{w}_0|F)$, analogous to Eq. (2.181). An equivalent expression for $\tilde{g}$ (Fano, 1961) is the principal value integral

$$\tilde{g}(r, r') = \frac{2}{\pi} P \int \frac{dE'}{E' - E} \tilde{w}_0(E'; r)\tilde{w}_0(E'; r'). \tag{3.35}$$

The formal solution of Eq. (3.28) is

$$f = \tilde{w}_0 - \tilde{G}(H - E_\alpha)\phi_\alpha c_\alpha. \tag{3.36}$$

This implies

$$(\alpha|H|f) = (\alpha|H|\tilde{w}_0) - (\alpha|(H - E_\alpha)\hat{G}(H - E_\alpha)|\alpha)c_\alpha \tag{3.37}$$

$$= (\alpha|H|\tilde{w}_0) + \Delta_\alpha c_\alpha, \tag{3.38}$$

where the orthogonality conditions are used to write Δ_α in a symmetric form. Equations (3.27) and (3.38) can be combined to give

$$c_\alpha = (E - E_\alpha - \Delta_\alpha)^{-1}(\alpha|H|\tilde{w}_0). \tag{3.39}$$

It follows from Eqs. (3.30) that $\tan \eta_{\mathrm{res}}$ is the asymptotic coefficient of $\tilde{w}_1(r)$ in $u(r)$ or $f(r)$. From Eq. (3.36) and the asymptotic form of $\hat{G}F$ in general this is

$$\tan \eta_{\mathrm{res}} = -2(\tilde{w}_0|H|\alpha)c_\alpha$$

$$= -2(\tilde{w}_0|H|\alpha)(E - E_\alpha - \Delta_\alpha)^{-1}(\alpha|H|\tilde{w}_0). \tag{3.40}$$

This verifies Eq. (3.21), with

$$\varepsilon = (E - E_{\mathrm{res}})/\tfrac{1}{2}\Gamma_\alpha, \tag{3.41}$$

where

$$E_{\text{res}} = E_\alpha + \Delta_\alpha \tag{3.42}$$

$$\Gamma_\alpha = 4(\tilde{w}_0|H|\alpha)^2. \tag{3.43}$$

The energy shift Δ_α and width Γ_α are both functions of E. An eigenvalue E_α of the Hamiltonian matrix H_{QQ} corresponds to a resonance only if Δ_α and Γ_α vary slowly near E_α and if Δ_α is small. In the present example of a potential well with an external barrier, there are obviously three ranges of energy with distinct behavior. If $E_\alpha < 0$, ϕ_α can be a true bound-state eigenfunction of H. Then the functions $\tilde{w}_0$ are also eigenfunctions, and the orthogonality condition is satisfied automatically. In this case, Δ_α is zero, the numerator of Eq. (3.40) vanishes, and $E - E_\alpha$ is greater than zero for E in the scattering continuum. Thus the scattering continuum is not directly perturbed by a bound state. If E is greater than zero but below the barrier energy $\bar{E}$, Eq. (3.40) can describe a resonance. The resonance energy is such that

$$E_{\text{res}} - E_\alpha - \Delta_\alpha(E_{\text{res}}) = 0 \tag{3.44}$$

and the width is $\Gamma_\alpha(E_{\text{res}})$.

If the basis set used for H_{QQ} were defined by a fixed boundary condition at the radius $\bar{r}$ corresponding to the potential maximum $\bar{E}$, the discrete set of eigenvalues $0 < E_\alpha < \bar{E}$ could all be identified with resonances. They all correspond to orthogonal wave functions that do not penetrate outside the potential barrier. However, a basis set of quadratically integrable functions extending outside the barrier can represent the true wave function for any finite r. With such basis sets, the number of eigenvalues in the range $0 < E_\alpha < \bar{E}$ increases indefinitely with the effective radius of completeness. Most eigenvalues of H_{QQ} computed in such a basis cannot be identified with resonances. In practice, to be so identified, an eigenvalue must be insensitive to changes in the basis, and the eigenvector must describe a function localized in the effective potential well defined by the target system. These properties are examined in the *stabilization method* (Taylor, 1970; Hazi and Taylor, 1970; Fels and Hazi, 1971, 1972; Hazi and Fels, 1971). This method will be described in detail in Section 3.4.

The derivation given above is an application of the Feshbach formalism to a shape resonance. The argument can be extended to describe multichannel resonances in general. Background eigenchannel states, orthogonal to the entire Hilbert space $\{\Phi_\mu\}$, are defined as solutions of Eq. (1.12), omitting the matrix optical potential due to a specified eigenvector $\{c_{\mu\alpha}\}$, energy E_α, of the bound–bound matrix $H_{\mu\nu}$, equivalent to H_{QQ}. For background eigenchannels σ with eigenphases η_σ and eigenchannel vector components $x_{p\sigma}$, the asymptotic radial channel orbitals are

$$\tilde{w}_{0p\sigma} \sim \tilde{w}_{0pp} x_{p\sigma} \cos \eta_\sigma + \tilde{w}_{1pp} x_{p\sigma} \sin \eta_\sigma,$$
$$\tilde{w}_{1p\sigma} \sim -\tilde{w}_{0pp} x_{p\sigma} \sin \eta_\sigma + \tilde{w}_{1pp} x_{p\sigma} \cos \eta_\sigma, \tag{3.45}$$

as in Eqs. (2.154) and (2.169).

The Green's function linear operator is a matrix

$$\sum_\sigma \tilde{G}^\sigma_{pq} \tag{3.46}$$

with a kernel $\tilde{g}^\sigma_{pq}(r, r')$ that is multiplied by appropriate angular and spin functions and vector-coupled and antisymmetrized into target state functions Θ_p and Θ_q. In the asymptotic region of large r or large r' the radial kernel function is

$$\tilde{g}^\sigma_{pq}(r, r') \sim 2\tilde{w}_{0p\sigma}(r_<)\tilde{w}_{1p\sigma}(r_>), \tag{3.47}$$

but additional orthogonality terms are present in the range of the assumed quadratically integrable orbital basis set. When interaction with state α is taken into account, the asymptotic radial open-channel functions become

$$f_{p\sigma}(r) \sim \tilde{w}_{0p\sigma} + \sum_{\sigma'} \tilde{w}_{1p\sigma'} \Delta K_{\sigma'\sigma}, \tag{3.48}$$

defining a perturbing reactance matrix ΔK as a generalization of $\tan \eta_{\text{res}}$. The argument leading to Eq. (3.40) follows as in the single-channel case, giving

$$\Delta K_{\sigma'\sigma} = -2 \sum_q M_{0q\sigma',\alpha}(E - E_\alpha - \Delta_\alpha)^{-1} \sum_p M_{\alpha,0p\sigma}, \tag{3.49}$$

where

$$M_{\alpha,0p\sigma} = (\Psi_\alpha | H - E_\alpha | \mathscr{A}\Theta_p Y_p \tilde{w}_{0p\sigma}) \tag{3.50}$$

is a bound–free matrix element in which Ψ_α is defined by the eigenvector $\{c_{\mu\alpha}\}$ of the bound–bound matrix $H_{\mu\nu}$. From Eq. (3.37), extended to the multichannel case, the energy width

$$\Delta_\alpha = -\sum_p \sum_q \sum_\sigma (\alpha | (H - E_\alpha)\tilde{G}^\sigma_{pq}(H - E_\alpha) | \alpha). \tag{3.51}$$

Since ΔK is a matrix, the appropriate generalization of Eq. (3.21) is a matrix expression. This can be obtained by defining a unit column vector with components

$$v_{\sigma\alpha} = 2\Gamma_\alpha^{-1/2} \sum_p M_{0p\sigma,\alpha}. \tag{3.52}$$

The total width Γ_α is defined by

$$\Gamma_\alpha = 4 \sum_\sigma \left(\sum_p M_{0p\sigma,\alpha} \right)^2, \tag{3.53}$$

as the normalization constant in Eq. (3.52). This is the direct generalization of Eq. (3.43). With these definitions, Eq. (3.49) becomes

$$\Delta K_{\sigma'\sigma} = -v_{\sigma'\alpha} \varepsilon_\alpha^{-1} v_{\alpha\sigma}^\dagger, \tag{3.54}$$

where

$$\varepsilon_\alpha = (E - E_\alpha - \Delta_\alpha)/\tfrac{1}{2}\Gamma_\alpha \tag{3.55}$$

$$= (E - E_{\text{res}})/\tfrac{1}{2}\Gamma_\alpha. \tag{3.56}$$

Since ΔK is defined in the basis of background eigenchannels, as defined by Eq. (3.46), it is convenient to transform the vectors v into another set of unit column vectors

$$y_{p\alpha} = \sum_\sigma x_{p\sigma} v_{\sigma\alpha}, \tag{3.57}$$

using the coefficients $x_{p\sigma}$ of the eigenchannel vectors. This converts Eqs. (3.45) and (3.48) to a representation in which the coefficients of the background functions $\tilde{w}_{0pp}$ and $\tilde{w}_{1pp}$ are real symmetric matrices of the form

$$(\cos \eta_0)_{pq} = \sum_\sigma x_{p\sigma} x_{q\sigma} \cos \eta_\sigma, \tag{3.58}$$

$$(\sin \eta_0)_{pq} = \sum_\sigma x_{p\sigma} x_{q\sigma} \sin \eta_\sigma. \tag{3.59}$$

From the general definition, Eq. (1.17), using this representation, the full K matrix is

$$K = (\sin \eta_0 - \varepsilon_\alpha^{-1} \cos \eta_0 y_\alpha y_\alpha^\dagger)(\cos \eta_0 + \varepsilon_\alpha^{-1} \sin \eta_0 y_\alpha y_\alpha^\dagger)^{-1}, \tag{3.60}$$

in a matrix notation. From its definition, Eq. (1.20), the S matrix is

$$S = (\cos \eta_0 + i \sin \eta_0)\left(I - \frac{i}{\varepsilon_\alpha} y_\alpha y_\alpha^\dagger \right)\left(I + \frac{i}{\varepsilon_\alpha} y_\alpha y_\alpha^\dagger \right)^{-1} (\cos \eta_0 - i \sin \eta_0)^{-1}$$

$$= \exp(i\eta_0)\left(I - \frac{2i}{\varepsilon_\alpha + i} y_\alpha y_\alpha^\dagger \right) \exp(i\eta_0). \tag{3.61}$$

If the background phase matrix η_0 and the total width and shift functions vary slowly near E_α and if Δ_α is small, this formula describes a multichannel resonance. These conditions can be satisfied only for nonoverlapping resonances, when the width is small compared with the spacing between adjacent resonances.

The general properties of multichannel resonances follow from Eq. (3.61), given by Brenig and Haag (1959). This equation shows that the analytically continued S matrix has a pole at $\varepsilon = -i$ in the vicinity of a resonance. If Γ and Δ are constant, the pole is at

$$E = E_\alpha + \Delta_\alpha - \tfrac{1}{2}i\Gamma_\alpha, \tag{3.62}$$

displaced below the physical E axis in the complex energy plane. It follows from Eq. (3.61) that a resonance affecting n coupled channels is described by the n eigenphase solutions $\eta_\tau(E)$ of the equation (Macek, 1970)

$$E - E_{\text{res}} = \frac{1}{2}\Gamma \sum_{\sigma=1}^{n} v_\sigma^2 \cot\left[\eta_\sigma(E) - \eta_\tau(E)\right], \tag{3.63}$$

where v_σ is a component of the unit vector, defined by Eq. (3.52), in the eigenchannel basis. If the background eigenphases η_σ and the vector components v_σ vary sufficiently slowly with energy, the derivative of Eq. (3.63) gives

$$1 = \frac{1}{2}\Gamma \frac{d\eta_\tau}{dE} \sum_{\sigma=1}^{n} v_\sigma^2 \csc^2\left[\eta_\sigma - \eta_\tau(E)\right]. \tag{3.64}$$

This implies that the eigenphases η_τ are nondecreasing functions of energy near a resonance. Typical behavior is shown for a three-channel example in Fig. 3.1 (Oberoi and Nesbet, 1973c). Because the eigenphases can be defined in terms of $\tan \eta_\tau$ as eigenvalues of the real symmetric K matrix, a noncrossing rule holds, as is evident in Fig. 3.1, but the whole pattern of eigenphases must shift upward, modulo π, in passing through the resonance.

The determinant of the S matrix is $\exp(2i\Sigma\eta)$, where $\Sigma\eta$ is the sum of eigenphases. The determinant of the middle term in Eq. (3.61) is

$$\det\left(I - \frac{2i}{\varepsilon + i} yy^\dagger\right) = \frac{\varepsilon - i}{\varepsilon + i} \tag{3.65}$$

$$= \exp\left[2i\eta_{\text{res}}(\varepsilon)\right], \tag{3.66}$$

defining a scalar resonance phase function $\eta_{\text{res}}(\varepsilon)$ that increases from 0 to π radians as ε varies from $-\infty$ to $+\infty$. If the background S matrix remains constant, its determinant is $\exp(2i\Sigma\eta_\sigma)$. It follows from Eqs. (3.61) and (3.66) that the sum of perturbed eigenphases, given by

$$\Sigma\eta_\tau(\varepsilon) = \Sigma\eta_\sigma + \eta_{\text{res}}(\varepsilon) \tag{3.67}$$

increases by π radians as ε traverses a resonance. Equation (3.67) shows that the sum of eigenphases, for a multichannel resonance, has much simpler behavior than the individual eigenphases. Since Eqs. (3.66) and (3.21) are equivalent, the sum of eigenphases varies with energy near a resonance

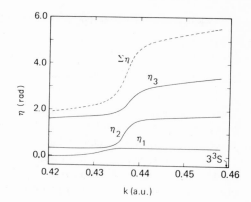

Figure 3.1. Eigenphases for a three-channel resonance (Oberoi and Nesbet, 1973c, Fig. 3).

exactly as does the phase shift for single-channel scattering. This is illustrated in Fig. 3.1.

The S matrix, as given by Eq. (3.61), can be written in the form

$$S = S_A + \exp{[2i\eta_{\text{res}}(\varepsilon)]}S_B, \tag{3.68}$$

where

$$S_A = \exp{(i\eta_0)}(I - yy^\dagger)\exp{(i\eta_0)}, \tag{3.69}$$

$$S_B = \exp{(i\eta_0)}yy^\dagger\exp{(i\eta_0)}, \tag{3.70}$$

and $\eta_{\text{res}}(\varepsilon)$ is defined by Eq. (3.66). The matrix S_A is of rank $n - 1$ and S_B is of rank 1. The background S matrix is

$$S_A + S_B = \exp{(2i\eta_0)}. \tag{3.71}$$

Since $\exp{(2i\eta_{\text{res}})}$ is a scalar phase factor, Eq. (3.68) implies that each matrix element S_{pq} describes a circle of phase $2\eta_{\text{res}}(\varepsilon)$ in the complex plane as ε increases, if the background S matrix can be treated as constant. At the resonance position, $\varepsilon = 0$, the S matrix is $S_A - S_B$. This behavior can be used to determine matrices S_A and S_B by interpolation, giving the background S matrix and the partial width amplitude vector y. The resonance energy E_{res} and width Γ are most easily determined directly from the sum of eigenphases.

A multichannel resonance search procedure for use with variational scattering calculations has been described by Nesbet and Lyons (1971). It is assumed that an eigenvalue E_α of H_{QQ} that meets the criteria of the stabilization method (Taylor, 1970; Taylor and Hazi, 1976) has been located in a preliminary calculation. If E_α corresponds to a resonance, the sum of eigenphases will be described by Eqs. (3.67) and (3.21) near E_α. If E is sufficiently close to E_{res}, a three-parameter formula can be used to fit the eigenphase sum. Expressed in terms of k values for single-channel scattering this is

$$\eta(k^2) = \eta_0 + \tan^{-1}[\Gamma/(k_{res}^2 - k^2)]. \tag{3.72}$$

For multichannel scattering, the same formula is used with η and η_0 replaced by $\Sigma\eta$ and $\Sigma\eta_0$, respectively. For each value of k that is considered, a multichannel variational calculation is carried out, determining eigenphases and eigenchannel vectors. The resonance search method will be described for the single-channel case.

Phase shifts η_1, η_2, η_3 obtained for three values of $k(k_1, k_2, k_3)$ can be used to determine the three parameters k_{res}, Γ, and η_0 in Eq. (3.72). Instead of η_0, it is somewhat more convenient to use the phase shift at resonance

$$\bar{\eta} = \eta_0 + \tfrac{1}{2}\pi, \tag{3.73}$$

as the third parameter. Equation (3.73) can be written in the form

$$\tan(\bar{\eta} - \eta_k) = (k_{res}^2 - k^2)/\Gamma, \tag{3.74}$$

where η_k denotes $\eta(k^2)$.

The multiple-angle formula for $\tan(\alpha - \beta)$ can be used to eliminate $\bar{\eta}$, giving

$$(k_2^2 - k_1^2)\cot(\eta_2 - \eta_1) = \Gamma + \Gamma^{-1}(k_{res}^2 - k_2^2)(k_{res}^2 - k_1^2),$$

$$(k_3^2 - k_2^2)\cot(\eta_3 - \eta_2) = \Gamma + \Gamma^{-1}(k_{res}^2 - k_3^2)(k_{res}^2 - k_2^2). \tag{3.75}$$

If Eq. (3.74) is used to replace $(k_{res}^2 - k_2^2)/\Gamma$ by $\tan(\bar{\eta} - \eta_2)$, k_{res}^2 can be eliminated by subtracting the two Eqs. (3.75) to give

$$\bar{\eta} = \eta_2 + \tan^{-1}\left[\frac{(k_2^2 - k_1^2)\cot(\eta_2 - \eta_1) - (k_3^2 - k_2^2)\cot(\eta_3 - \eta_2)}{k_3^2 - k_1^2}\right]. \tag{3.76}$$

If this value is used to compute $\tan(\bar{\eta} - \eta_1)$ and $\tan(\bar{\eta} - \eta_3)$, Eq. (3.74) gives formulas for Γ and k_{res}^2,

$$\Gamma = \frac{(k_3^2 - k_1^2)}{[\tan(\bar{\eta} - \eta_1) - \tan(\bar{\eta} - \eta_3)]}, \tag{3.77}$$

$$k_{res}^2 = \frac{[k_3^2\tan(\bar{\eta} - \eta_1) - k_1^2\tan(\bar{\eta} - \eta_3)]}{[\tan(\bar{\eta} - \eta_1) - \tan(\bar{\eta} - \eta_3)]}. \tag{3.78}$$

Following the analysis given in Section 2.3, the variational method provides a crude estimate of the zeroes of $\tan\eta$ and $\cot\eta$ lying closest to E_α (Nesbet and Lyons, 1971). Since these two points are separated by a phase interval of $\pi/2$ radians, their relative separation fixes the magnitude of the interval of E or k to be scanned to locate a resonance near E_α. The three-point formula given by Eqs. (3.76)–(3.78) is used iteratively, starting with k_1 and k_3 taken to be the estimated zeroes of $\tan \eta$ or $\cot \eta$, and k_2 to be their geometric mean. At each iteration, the computed k_{res} is compared with the current k_1, k_2, and k_3. If Δk is the smallest value of $|k_{\mathrm{res}} - k_i|$, the k values for the next iteration are taken to be $k_{\mathrm{res}} - \Delta k$, k_{res}, and $k_{\mathrm{res}} + \Delta k$, requiring only two new phase-shift calculations. The search terminates when Δk is smaller than some specified criterion value.

If there is no resonance in the scattering channels determined by the wave function corresponding to E_α, experience with this search procedure indicates that it diverges rapidly, and can be terminated by testing Δk against an upper limit. This procedure defines the resonance energy as the point of steepest rise of $\Sigma \eta$ and determines the resonance parameters $\Sigma \eta_0$ and Γ there. This has the practical advantage of not attempting to fit a rather ill-defined background phase variation, but it may require a correction for such variation if Γ is large or if other structural features are close in energy.

Figure 3.2 illustrates this procedure for a two-channel problem (Nesbet and Lyons, 1971). The points shown in the figure are k_α, corresponding to

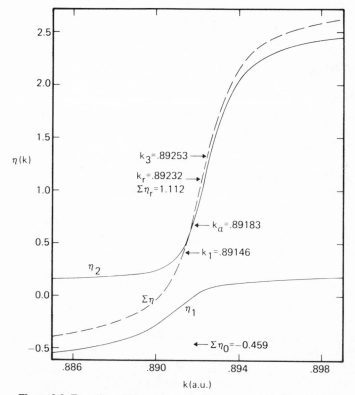

Figure 3.2. Two-channel resonance search (Lyons and Nesbet, 1971, Fig. 4).

the energy eigenvalue E_α, k_1, and k_3, the estimated zeroes of $\tan \Sigma\eta$ and $\cot \Sigma\eta$, respectively, and $k_r = k_{res}$ as determined by the resonance search. Eigenphases η_1 and η_2 are shown together with their sum, which is used in the search procedure. The computed resonance parameters are

$$\Sigma\bar{\eta} = 1.112 \text{ radians},$$

$$\Gamma = 4.947 \times 10^{-2} \text{ eV}, \qquad (3.79)$$

$$k_{res} = 0.89232 a_0^{-1}.$$

A satisfactory fit to the computed eigenphases η_1 and η_2 is obtained with Eq. (3.63) if

$$v_{12}^2 = 0.8819,$$
$$\qquad (3.80)$$
$$v_{21}^2 = 0.1181.$$

Here the double subscript (12) refers to the background eigenphase approached by η_1 for $k < k_{res}$ and by η_2 for $k > k_{res}$, and similarly for subscript (21). The background eigenphases are assumed to be the constants $\eta_{12}^0 = -0.62533$ and $\eta_{21}^0 = 0.16548$.

3.3. Scattering Near Thresholds

In electron-impact excitation, the inelastically scattered electron has energy that goes to zero at the excitation threshold. The very low velocity implies a long delay time, which is related to energy-dependent structure in scattering cross sections by Eqs. (3.8) and (3.9).

The single-channel threshold theory, which will be presented first, makes use of the analytic theory of singularities of the S matrix, or $\exp(2i\eta)$ for a single channel. The formal mathematical theory is helpful in interpreting structural features of cross sections near thresholds. Geltman (1969) summarizes the theory of resonance and threshold structures. The discussion here follows McVoy (1967) except for the treatment of long-range potentials.

At very low energies, scattering is dominated by the longest range part of the effective potential function. The effects of long-range potentials on elastic cross sections near threshold were studied by O'Malley et al. (1961) and Levy and Keller (1963). This work was extended to multichannel scattering by Bardsley and Nesbet (1973). For $l > 0$, if the scattering potential vanishes more rapidly than r^{-2}, for large r, the leading asymptotic potential term determines the threshold behavior of phase shifts and cross sections. More difficult analysis is required for the special cases of scattering by a static electric dipole moment (Gailitis and Damburg, 1963b) and of

Coulomb scattering, which will not be considered here. These cases are discussed by Burke (1977).

In multichannel scattering, the single-channel theory applies to elastic scattering from each target state just above its excitation threshold. Resonances or virtual states in the elastic scattering continuum of a particular target state are described by this theory, as well as one-electron bound states in the effective potential appropriate to this continuum. However, a state that is bound relative to an excited parent state in general lies in the scattering continuum of lower target states and is actually a CE1 resonance as described by Taylor (1970). Since Eq. (3.65) generalizes Eq. (3.15) to the case of nonoverlapping multichannel resonances, Eq. (3.61) implies that all elements of the analytically continued S matrix have coincident poles at $E_{res} - \frac{1}{2}i\Gamma$. The analytic theory of these singularities applies to the multichannel case, but it is most easily presented for single-channel scattering.

Following McVoy (1967), the S matrix for single-channel scattering can be expressed as a function of complex k

$$S(k) = \exp[2i\eta(k)] = f(-k)/f(k). \tag{3.81}$$

Here $f(k)$ is a *Jost function* whose zeroes determine poles of $S(k)$. The upper half-plane of complex k is mapped onto the entire complex energy plane, with a branch cut on the positive real E axis. The lower half-plane of k maps onto a second "nonphysical" sheet of the Riemann surface of complex E. Zeroes of $f(k)$ in the upper half-plane can occur only on the positive imaginary axis at points

$$k_b = i\beta, \qquad \beta > 0, \tag{3.82}$$

and correspond to bound states. Zeroes of $f(k)$ in the lower half-plane occur either as pairs of roots

$$\begin{aligned} k_\rho &= \alpha - i\beta, \\ -k_\rho^* &= -\alpha - i\beta, \qquad \alpha, \beta > 0, \end{aligned} \tag{3.83}$$

corresponding to a resonance pole of S or as single points on the negative imaginary axis

$$k_v = -i\beta, \qquad \beta > 0, \tag{3.84}$$

corresponding to virtual states.

At a resonance, S contains the factor

$$\frac{1 + i\tan\eta_\rho}{1 - i\tan\eta_\rho} = \frac{(-k - k_\rho)(-k + k_\rho^*)}{(k - k_\rho)(k + k_\rho^*)}. \tag{3.85}$$

This implies

$$\tan \eta_\rho = \frac{-2k\beta}{k^2 - (\alpha^2 + \beta^2)} \tag{3.86}$$

so that, comparing with Eq. (3.11), the resonance parameters are

$$E_{\text{res}} = \tfrac{1}{2}k_{\text{res}}^2 = \tfrac{1}{2}(\alpha^2 + \beta^2), \tag{3.87}$$

$$\Gamma = 2k_{\text{res}}\beta. \tag{3.88}$$

If the scattering potential is varied so that a resonance approaches the threshold, $k_\rho \to 0$, β must vanish at least as rapidly as k_{res}, so Γ vanishes at least quadratically with k_{res}. For a resonance due to the centrifugal barrier in a partial wave with $l > 0$, β varies as k_{res}^{2l}, so Γ vanishes as k_{res}^{2l+1} if a model potential is varied to move a resonance toward threshold (McVoy, 1967). This analysis shows that shape resonances, which might otherwise be quite broad, can become very narrow if they occur in the threshold region, essentially because the centrifugal potential barrier becomes very wide if k_{res} is small. The two poles k_ρ and $-k_\rho^*$ of $S(k)$ approach the origin tangentially to the real k axis. Model studies show that for $l > 0$ the paired resonance poles can be pictured as colliding at the origin then separating as paired bound- and virtual-state poles on the positive and negative imaginary k axis, respectively.

If $l = 0$, there is no centrifugal barrier. As an attractive potential function is increased in strength, a zero of $f(k)$ approaches $k = 0$ as a single point on the negative imaginary axis. From Eq. (3.84), this is a virtual-state pole of the S matrix, which passes directly through the origin $k = 0$ to become a bound-state pole. Thus a virtual state is a phenomenon associated with s-wave scattering at a threshold. The phase shift for a virtual state corresponds to an S-matrix factor of the form

$$\frac{1 + i \tan \eta_v}{1 - i \tan \eta_v} = \frac{-k - k_v}{k - k_v} \tag{3.89}$$

or, from Eq. (3.84),

$$\tan \eta_v = k/\beta. \tag{3.90}$$

If β is small, the phase shift rises rapidly from threshold, but the total increase is limited to $\pi/2$ radians. For s-wave elastic scattering, the partial cross section given by Eq. (1.27) is

$$\sigma = 4\pi/(k^2 + \beta^2), \tag{3.91}$$

which has a finite nonzero value at threshold.

Similarly, for a bound state at energy $-\beta^2/2$, just below a threshold, the phase shift varies as

$$\tan \eta_b = -k/\beta, \tag{3.92}$$

from Eq. (3.82). The phase shift descends linearly with k from a value at threshold that can be taken to be π radians, representing the contribution of the bound state as a resonance of zero width. The partial cross section is again given by Eq. (3.91), finite at threshold. Thus the scattering effect of a bound state at energy $-\beta^2/2$ is the same as that of a virtual state at the *same* energy, but displaced onto the nonphysical sheet of the complex E plane. Since the nearest physical energy value is the threshold itself, a virtual state appears physically as a threshold effect. Real and virtual states are easily distinguished in theory because of the opposite sign of the slope of $\tan \eta$ at threshold in the two cases. Because of their similar effect on elastic scattering cross sections, they cannot be distinguished by a scattering experiment at energies above the relevant threshold, unless a phase-shift analysis is possible.

When extended to multichannel scattering, these analytic properties of $S(k)$ apply to the eigenphase associated with the new open channel at an excitation threshold. Elastic scattering in the new channel is affected by a resonance below threshold as if it were a bound state, giving a finite threshold cross section. A virtual state has a similar effect.

The multichannel theory of threshold effects can be based on the Lippmann–Schwinger equation, Eq. (2.206), which provides an exact solution of the multichannel scattering equations given by Eq. (2.199). If the model Green's function refers to free-particle scattering, it is a sum of single-channel terms. The kernel in channel α is

$$g_\alpha(r, r') = 2w_{0\alpha}(r_<)w_{1\alpha}(r_>). \tag{3.93}$$

The Lippmann–Schwinger equation is

$$u_{\alpha\beta}(r) = w_{0\alpha}(r)\delta_{\alpha\beta} - \sum_\gamma g_\alpha(r, r')V_{\alpha\gamma}(r')u_{\gamma\beta}(r')\,dr', \tag{3.94}$$

or, in a matrix notation,

$$u = w - GVu. \tag{3.95}$$

For single-channel potentials that vanish more strongly than r^{-2} at large r, the free-particle functions w_0 and w_1 are given in terms of spherical Bessel functions

$$w_{0\alpha}(r) = k_\alpha^{1/2} r j_{l\alpha}(k_\alpha r),$$
$$w_{1\alpha}(r) = -k_\alpha^{1/2} r n_{l\alpha}(k_\alpha r), \tag{3.96}$$

in open-channel α. The asymptotic form of Eq. (3.94) gives the K matrix as

$$K_{\alpha\beta} = -2 \sum_{\gamma} \int_0^{\infty} w_{0\alpha}(r) V_{\alpha\gamma}(r) u_{\gamma\beta}(r) \, dr. \tag{3.97}$$

In certain limiting cases the second term in Eq. (3.95) can be neglected, and Eq. (3.97) can be approximated by

$$K_{\alpha\beta} \cong -2 \int_0^{\infty} w_{0\alpha}(r) V_{\alpha\beta}(r) w_{0\beta}(r) \, dr. \tag{3.98}$$

This is the first term of the Born series for $K_{\alpha\beta}$ (Joachain, 1975), obtained by expanding in powers of a parameter that multiplies the matrix potential function $V_{\alpha\beta}$. In general, the first Born approximation is valid if all elements of the K matrix are small since this implies that the second term of Eq. (3.95) is small. This is true in the high-energy limit, which provides the principal practical use of this approximation, but it is also true for K-matrix elements in the threshold region for one or both of the channel indices. For potentials weaker than r^{-2} at large r, these matrix elements vanish at threshold.

As shown by Levy and Keller (1963) for single-channel scattering, the K-matrix Born approximation can be used to analyze the analytic behavior of such K-matrix elements near threshold. The multichannel analysis here will follow the argument of Bardsley and Nesbet (1973).

At a given excited-state threshold, there will be one or more old channels, open below the threshold, which can be labeled by $\alpha, \beta, \ldots$. New channels opening at the threshold will be labeled by $p, q, \ldots$. The new channels have a common k value

$$k_p = k_q = \cdots = k.$$

Threshold behavior is determined by the dependence of matrix elements K_{pq} and $K_{\alpha p}$ on k, in the limit $k \to 0$.

The regular free-particle functions $w_{0p}(r)$ increase from the origin as some power of kr but are normalized so that they oscillate with magnitude $k^{-1/2}$ at large r. As k decreases, these functions are negligible in an expanding range of r about the origin. For $l_p > 0$, this is defined by a radius corresponding to the classical turning point for a particle moving in a centrifugal potential. For any short-range potential $V_{pq}(r)$, explicit expansion of Eq. (3.98) in powers of k gives

$$K_{pq}^S = -k^{l_p+l_q+1} \left[\int_0^{\infty} 2 V_{pq} r^{l_p+l_q+2} \, dr + O(k^2) \right], \tag{3.99}$$

from the functional forms of the regular spherical Bessel functions. This gives an explicit threshold law for elastic scattering in the new channels. If the interaction potential contains a long-range component of the form Cr^{-s}, there is a contribution to K_{pq} from the integral

$$K_{pq}^L = -2Ck \int_0^\infty r^{2-s} j_{lp}(k_p r) j_{lq}(k_q r)\, dr, \qquad (3.100)$$

which may dominate the short-range term. If $s < (l_p + l_q + 3)$, this integral is well defined and gives [Abramowitz and Stegun, 1964, Eqs. (11.4.33) and (15.1.20)]

$$K_{pq}^L = \frac{-\pi C}{2^{s-1}} k^{s-2} \frac{\Gamma(s-1)\Gamma[\frac{1}{2}(l_p + l_q - s + 3)]}{\Gamma[\frac{1}{2}(l_p - l_q + s)]\Gamma[\frac{1}{2}(l_p + l_q + s + 1)]\Gamma[\frac{1}{2}(l_q - l_p + s)]}. \qquad (3.101)$$

When $s < (l_p + l_q + 3)$ then $(s - 2) < (l_p + l_q + 1)$ and the term k^{s-2} is dominant for small k. However, if $(s - 2) \geq (l_p + l_q + 1)$, the integrand of Eq. (3.100) must be modified to avoid a singularity at the origin as it would be for any physical potential function. In this case the dominant power of k is $k^{l_p + l_q + 1}$ since the integral is made finite by the short-range behavior of the potential. Since $s > 2$ by assumption, K_{pq} always vanishes at threshold, so it is consistent to use the Born approximation for this analysis.

As an example of this threshold law, consider elastic scattering by a polarization potential, given by $-\alpha_d/2r^4$ at large r if α_d is the electric dipole polarizability of the target state. Then $s = 4$, and $s - 2$ is less than $2l + 1$ if $l > 0$ for elastic scattering. The threshold law implies that the leading term in $\tan \eta_l$ is proportional to k^2 for small k for all partial waves with $l > 0$, but the s-wave phase shift near threshold is determined by the short-range structure of the effective scattering potential. In this example Eq. (3.101) gives the very useful formula

$$\tan \eta_l \cong \frac{\pi \alpha_d k^2}{(2l + 3)(2l + 1)(2l - 1)}, \qquad l > 0. \qquad (3.102)$$

Detailed analysis (O'Malley, 1963) of the effect of the long-range polarization potential gives a modified effective range formula for $l = 0$,

$$\tan \eta_0 \cong -kA_0(1 + \tfrac{4}{3}\alpha_d k^2 \ln k) - \tfrac{1}{3}\pi \alpha_d k^2, \qquad (3.103)$$

where A_0 is the scattering length, defined as the limit of $-k^{-1} \tan \eta_0$ for $k \to 0$. It should be noted that the final term in Eq. (3.103) is just equal to Eq. (3.102) for $l = 0$, but the terms involving the scattering length dominate $\tan \eta_0$ for small k. For an approximation consistent with Eq. (3.103), a short-range term proportional to k^3 must be included together with the k^2 term of Eq. (3.102) for the p-wave phase shift, $l = 1$ (O'Malley, 1963).

The threshold theory for inelastic scattering requires consideration of K-matrix elements $K_{\alpha p}$, where index α refers to a channel open below threshold. While it cannot be assumed that $u_{\alpha\beta}(r)$ is adequately approximated by $w_{0\alpha}(r)\delta_{\alpha\beta}$, the symmetry of the K matrix can be used to justify use of the Born formula, Eq. (3.98), for $K_{\alpha p}$ if the second index refers to a new channel that becomes open at the threshold $k = 0$. For any short-range interaction potential, expansion of $w_{0p}(r)$ in powers of kr gives a result similar to Eq. (3.99), with a term proportional to $k^{l_p+1/2}$ dominant for small k. When $V_{\alpha p}$ has a long-range component Cr^{-s}, the contribution to Eq. (3.98) is

$$
\begin{aligned}
K_{\alpha p}^L &= -2Ck_{\alpha}^{1/2}k^{1/2}\int_0^{\infty} r^{2-s}j_{l\alpha}(k_{\alpha}r)j_{lp}(kr)\,dr \\
&= -\frac{\pi C}{2^{s-1}}k_{\alpha}^{s-l_p-5/2}k^{l_p+1/2}\frac{\Gamma[\frac{1}{2}(l_{\alpha}+l_p-s+3)]}{\Gamma(l_p+\frac{3}{2})\Gamma[\frac{1}{2}(l_{\alpha}-l_p+s)]}\left[1+O\!\left(\frac{k^2}{k_{\alpha}^2}\right)\right],
\end{aligned}
$$

(3.104)

valid for $s < (l_{\alpha} + l_p + 3)$. For small k, this has the same behavior as $K_{\alpha p}^S$. If $s \geq (l_{\alpha} + l_p + 3)$, the physical potential must be modified by short-range terms that make the integral finite, so the short-range behavior dominates. It can be concluded that long-range interactions do not change the threshold law for K-matrix elements connecting old and new channels, which vary as the $l_p + \frac{1}{2}$ power of k in all cases. This implies that such nondiagonal matrix elements start from zero at threshold. For electron scattering by neutral atoms, this in turn implies that excitation cross sections start from zero values at their thresholds. However, for an s wave in the new channel, $l_p = 0$ and the excitation cross section varies as $(\Delta E)^{1/2}$, which rises with vertical slope at threshold.

For energies near a new threshold energy, if $l_p = 0$ in a new channel, the abrupt rise from zero of the excitation cross section can influence scattering cross sections in the old channels. Because matrix elements $K_{\alpha p}$ vary as $k^{l_p+1/2}$ near threshold, the resulting S-matrix elements have square-root branch points as functions of k. This leads to characteristic structure in the form of cusps or rounded steps in cross sections for processes involving only the old channels (Wigner, 1948). The theory of these multichannel threshold effects has been summarized by Brandsden (1970) and Nesbet (1975b).

For orbital s-waves in n channels opening at a given threshold, if M channels are open below the threshold and all relevant channels are coupled by a multichannel potential function, the open-channel K matrix changes from dimension $M \times M$ to dimension $N \times N$ at the threshold, where $N = M + n$. Below threshold, this matrix will be denoted by $\bar{K}^{MM}$. Above threshold it is of the form

$$K^{NN} = \begin{pmatrix} K^{MM} & K^{Mn} \\ K^{nM} & K^{nn} \end{pmatrix}. \tag{3.105}$$

Because $\bar{K}^{MM}$ contains the physical effects of the n channels that become closed below threshold, it differs from the direct analytic continuation of K^{MM}, which is independent of these channels. However, in the formal theory of multichannel scattering, the S matrix is defined so that the corresponding submatrix S^{MM} can be continued analytically through a threshold. The K matrix is defined, for open channels only, such that the open-channel submatrix of the generalized S matrix is given by

$$S = (I + iK)(I - iK)^{-1}, \tag{3.106}$$

or

$$(I + iT)(I - iK) = I, \tag{3.107}$$

where

$$T = (S - I)/2i. \tag{3.108}$$

It can be shown (Brandsden, 1970, pp. 152–160; Dalitz, 1961) that the correct K matrix below threshold is given by analytic continuation of the matrix defined above threshold by

$$\bar{K}^{MM} = K^{MM} + iK^{Mn}(I - iK^{nn})^{-1}K^{nM}. \tag{3.109}$$

Because the elements of K^{Mn} vanish at threshold, $\bar{K}^{MM}$ is continuous with K^{MM}, but the matrix elements are discontinuous in slope as functions of k or E if they refer to s-wave channels above threshold. The resulting cross sections are also discontinuous in slope, showing the cusp or step structures described by Wigner (1948).

A simple proof of Eq. (3.109) follows from Eq. (3.107). The $N \times N$ matrix $(I + iT)$ can be analytically continued through the threshold. Then the inverse of the open-channel submatrix defines $I - iK$ below threshold. This is

$$I - i\bar{K}^{MM} = [(I + iT)^{MM}]^{-1}$$
$$= \{[(I - iK^{NN})^{-1}]^{MM}\}^{-1}, \tag{3.110}$$

which relates the correct K matrix below threshold, $\bar{K}^{MM}$, to the analytic continuation of the K matrix above threshold, K^{NN}. In general, for a matrix A^{NN} partitioned as in Eq. (3.105), the $M \times M$ submatrix of the $N \times N$ inverse is given by

$$(A^{-1})^{MM} = [A^{MM} - A^{Mn}(A^{nn})^{-1}A^{nM}]^{-1} \tag{3.111}$$

unless A^{nn} is singular. When applied to Eq. (3.110) this implies

$$I - i\bar{K}^{MM} = I - iK^{MM} + K^{Mn}(I - iK^{nn})^{-1}K^{nM}, \tag{3.112}$$

which reduces to Eq. (3.109).

For orbital s-waves in the n channels opening at threshold, the submatrices of K^{NN} can be expressed in the form

$$K^{MM} = A^{MM},$$

$$K^{Mn} = (K^{nM})^{\dagger} = A^{Mn}k^{1/2}, \tag{3.113}$$

$$K^{nn} = A^{nn}k.$$

The matrices A vary slowly with k and are continuous with continuous first derivatives. They can be considered to be real and constant near the threshold.

Below threshold, k becomes $i\kappa$, with $\kappa > 0$. From Eq. (3.109),

$$\bar{K}^{MM} = A^{MM} - A^{Mn}A^{nM}\kappa + O(\kappa^2). \tag{3.114}$$

This formula can be used to compute scattering cross sections just below a threshold. Analytic continuation is not expected to be reliable beyond the linear region in κ because long-range potentials imply irregular dependence of the matrices A on the variable k beyond the leading terms indicated in Eqs. (3.113). For example, continuation of the term $k^3 \ln k$ in Eq. (3.103) below threshold requires more detailed analysis.

While the K matrix provides the most compact representation of threshold effects, it is generally more convenient to represent scattering data in terms of eigenphases η_σ and eigenchannel vectors $x_{p\sigma}$. For Eqs. (3.113) to be valid, certain conditions must be satisfied by analytic forms chosen to fit the vectors x. In particular, above threshold,

$$x^{MM} = \alpha^{MM}, \qquad x^{Mn} = \alpha^{Mn}k^{1/2}, \qquad x^{nM} = \alpha^{nM}k^{1/2}, \tag{3.115}$$

where the matrix of eigenvectors is not symmetric. The matrices α may contain terms linear in k. Since K can be constructed from its eigenvalues and eigenvectors in the form

$$K_{pq} = \sum_\sigma x_{p\sigma}x_{q\sigma} \tan \eta_\sigma, \tag{3.116}$$

the first of Eqs. (3.113) will be free of terms linear in k only if

$$\frac{\partial}{\partial k} K^{MM}_{pq}\bigg|_{k=0} = 0$$

$$= \sum_\sigma \left[\left(x_{p\sigma}\frac{\partial x_{q\sigma}}{\partial k} + \frac{\partial x_{p\sigma}}{\partial k}x_{q\sigma} \right) \tan \eta_\sigma + \frac{x_{p\sigma}x_{q\sigma}}{\cos^2 \eta_\sigma}\frac{\partial \eta_\sigma}{\partial k} \right]\bigg|_{k=0}. \tag{3.117}$$

Together with the orthogonality conditions

$$\sum_{\sigma} x_{p\sigma} x_{q\sigma} = \delta_{pq}, \tag{3.118}$$

Eqs. (3.117) determine the terms linear in k in x^{MM} and η_σ^M, if α^{Mn}, α^{nM}, and the constant leading terms in x^{MM} and η_σ^M are given.

This analysis leads to well-known simple formulas when $M = 1, N = 2$. Let eigenphase η_a correspond to the open channel below threshold, taking the value η at threshold. The second eigenphase η_b is proportional to k just above threshold. The eigenvectors are defined by a mixing coefficient χ such that

$$x_{2a} = \chi k^{1/2}, \qquad x_{1b} = -\chi k^{1/2}. \tag{3.119}$$

Sufficiently near threshold for χ to be assumed constant, Eqs. (3.114) and (3.117) imply

$$\eta_a = \begin{cases} \eta - \kappa \chi^2 \sin^2 \eta + O(\kappa^2), & k^2 < 0, \\ \eta + k\chi^2 \sin \eta \cos \eta + O(k^2), & k^2 > 0. \end{cases} \tag{3.120}$$

If the channel open below threshold is denoted by index 1 and the new channel is denoted by index 2, the elastic cross section below threshold is

$$\sigma_{11} = \frac{4\pi}{k_1^2} \sin^2 (\eta - \kappa \chi^2 \sin^2 \eta) + O(\kappa^2)$$

$$= \frac{4\pi}{k_1^2} \sin^2 \eta [1 - 2\kappa \chi^2 \cos \eta \sin \eta + O(\kappa^2)]. \tag{3.121}$$

Above threshold, this cross section becomes

$$\sigma_{11} = \frac{4\pi}{k_1^2} |x_{1a}^2 \exp (i\eta_a) \sin \eta_a + x_{1b}^2 \exp (i\eta_b) \sin \eta_b|^2$$

$$= \frac{4\pi}{k_1^2} [(1 - k\chi^2)^2 \sin^2 (\eta + k\chi^2 \sin \eta \cos \eta) + O(k^2)]$$

$$= \frac{4\pi}{k_1^2} \sin^2 \eta [1 - 2k\chi^2 \sin^2 \eta + O(k^2)]. \tag{3.122}$$

Equations (3.121) and (3.122) show the discontinuity in slope at $k = 0$ that characterizes a Wigner cusp or rounded step in an elastic cross section at an inelastic threshold.

The excitation cross section just above threshold in the present case is

$$\sigma_{12} = \frac{4\pi}{k_1^2} |x_{1a}x_{2a} \exp(i\eta_a) \sin \eta_a + x_{1b}x_{2b} \exp(i\eta_b) \sin \eta_b|^2$$

$$= \frac{4\pi}{k_1^2} k\chi^2 (1 - k\chi^2) \sin^2(\eta_a - \eta_b)$$

$$= \frac{4\pi}{k_1^2} \sin^2 \eta [k\chi^2 + O(k^2)]. \tag{3.123}$$

The total cross section above threshold is

$$\sigma_{11} + \sigma_{12} = \frac{4\pi}{k_1^2} \sin^2 \eta [1 + k\chi^2 \cos 2\eta + O(k^2)]. \tag{3.124}$$

These formulas show that the total cross section, like the elastic cross section alone, has discontinuous slope at the threshold. Thus it is not correct to attribute the Wigner cusp structure simply to a requirement of continuity of scattering flux at threshold. Equation (3.123) shows that the excitation cross section rises from threshold as $(\Delta E)^{1/2}$. There is no limit on the magnitude of χ.

The analysis given here makes it possible to deduce the K matrix just below an excitation threshold from the K matrix computed as a function of energy just above the threshold. This is useful in calculations by the matrix variational method. Since this method includes explicit continuum basis functions for the open channels with small k values above a threshold, it can be highly accurate. It is more difficult to include exponential basis functions with very small exponential parameters, required in principle for accurate variational calculations just below a threshold. Equation (3.109) avoids this problem by making it possible to extrapolate the K matrix through the threshold.

3.4. The Stabilization Method

In the stabilization method (Taylor, 1970; Hazi and Taylor, 1970), resonances are located by examining eigenvalues of the $(N + 1)$-electron Hamiltonian matrix in a quadratically integrable basis representation. Efficient methods developed for bound states can be used since the scattering continuum is not represented explicitly. This method is especially valuable in considering electron–molecule scattering resonances because of the inherent difficulty of full scattering calculations. As an adjunct to the matrix variational method, the stabilization method provides an efficient preliminary search for possible resonances, which can be characterized subsequently by full scattering calculations using the resonance search procedure described in Section 3.2 (Nesbet and Lyons, 1971).

In using the stabilization method, variational wave functions are constructed that have a structure anticipated on physical grounds, usually described by a single electronic configuration or by a small set of strongly interacting configurations. Then a matrix Hartree–Fock calculation is carried out using quadratically integrable basis functions. If no resonance exists of the assumed structure, it is found in practice that the Hartree–Fock calculation will not converge to a well-defined function (Taylor and Hazi, 1976). The basis set is modified within a range of auxiliary parameters for which convergence occurs, and calculations are extended to full configuration interaction calculations. Wave functions are selected by a criterion of maximum overlap with the matrix Hartree–Fock wave function so that the model of a localized $(N + 1)$-electron function of definite structure is maintained. The range of energy eigenvalues that correspond to wave functions of the specified inner structure in these calculations gives a rough estimate of the resonance width. If several eigenfunctions, with energy eigenvalues stabilized with respect to increase or modification of the variational basis, occur within this approximate width, the one having the greatest overlap with the corresponding matrix Hartree–Fock function is assumed to give the best estimate of the resonance energy and of the inner part of the true continuum wave function at this energy. In order to verify the identification of a true resonance, the stabilization calculation must be supplemented by a calculation of the resonance width.

Properties of this method were considered in model calculations by Hazi and Taylor (1970). The single-channel model potential contained an inner harmonic-oscillator well and a finite potential barrier. Oscillator eigenfunctions were used as a basis set. For basis sets of varying dimension (up to $N = 50$), functions computed from eigenvectors of the Hamiltonian matrix were compared to continuum eigenfunctions obtained by numerical integration. Figure 3.3 shows such a comparison. The integrated continuum function is scaled to agree in magnitude with the variational function at its first maximum. The expansion length is $N = 50$, and the continuum eigenfunction is computed at the same energy as the computed eigenvalue of the Hamiltonian matrix. The behavior shown in Fig. 3.3 is characteristic of this method. A very accurate representation of a true eigenfunction is obtained within a finite range defined by the basis set, outside of which the variational function goes smoothly to zero while the true continuum function oscillates with finite amplitude.

In this model calculation, the outer potential barrier peak is near $x = 2.0$. Hence the variational basis, as indicated in Fig. 3.3, is effectively

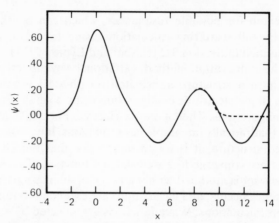

Figure 3.3. Comparison of exact and approximate wave functions (Hazi and Taylor, 1970, Fig. 3).

complete well into the external scattering region. The Feshbach resonance theory, as described in Section 3.2 would clearly give accurate values of the resonance parameters if the stabilized function of this example were used as the normalized function ϕ_α considered there and continuum basis functions $\tilde{w}_0$ and $\tilde{w}_1$ were evaluated only outside the potential barrier. This is the wave-function structure used in the matrix variational method. The resonance theory shows that the coefficient c_α of ϕ_α in the true wave function, given by Eq. (3.39), has a pole at E_{res}, which contains the energy shift. This indicates that the stabilized matrix eigenfunction, chosen by the criterion of maximum weight in the inner potential region, should have an energy eigenvalue close to E_{res}. As more basis functions are included, the residual energy shift should go to zero, but the width approaches the true resonance width as a finite limit.

The behavior of a stabilized energy eigenvalue for this model problem as a function of basis size is shown in Fig. 3.4 (Hazi and Taylor, 1970). A true resonance occurs at energy $E_{res} = 0.4413$, in dimensionless units. The stable behavior of eigenvalues near E_{res} is evident from the figure. However, the nth eigenvalue must decrease monotonically with expansion length N (Hylleraas and Undheim, 1930). Since there is no limit on the number of eigenvalues that must eventually lie below E_{res} in the scattering continuum, as N is increased there must be an infinite sequence of root crossings of the kind illustrated in Fig. 3.4. Although the residual energy shift cannot vanish in the immediate vicinity of such a crossing, a horizontal line connecting the stabilized portions of successive branches of the eigenvalue graph gives an accurate estimate of E_{res}. The slope of each branch in the stabilized region can be shown to be approximately proportional to the resonance width (Hazi and Taylor, 1970).

The resonance width Γ_α and residual energy shift Δ_α for a stabilized function ϕ_α with energy eigenvalue E_α depend on the interaction with the background scattering continuum. The formulas for these quantities, Eqs. (3.43) and (3.38), respectively, are dependent even for single-channel scattering on a third parameter, the background phase shift η_0. If two or more stabilized states associated with the same resonance are available, perhaps corresponding to different basis expansions, Hazi and Fels (1971) show that the resonance parameters (E_{res}, η_0, Γ) can be obtained from ϕ_α and the bound–free matrix elements

$$M_{\alpha S} = (\alpha|H - E_\alpha|w_0),$$
$$M_{\alpha C} = (\alpha|H - E_\alpha|w_1),$$
(3.125)

where w_0 and w_1 have asymptotic forms given by the first and second of Eqs. (2.1), respectively, and are regular at the coordinate origin. Since E_α is an eigenvalue of the bound–bound matrix, Eqs. (2.5) of the matrix variational method have finite solutions at E_α only if

$$M_{\alpha S} + M_{\alpha C} \tan \eta = 0.$$
(3.126)

This condition is used to determine the phase shift at eigenvalues E_α in the single-channel variational method of Harris (1967). If E_α is associated with

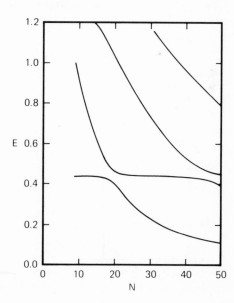

Figure 3.4. Behavior of the four lowest eigenvalues of a model problem (Hazi and Taylor, 1970, Fig. 5).

a resonance at E_{res}, Eqs. (3.126) and (3.20) give

$$\eta(E_\alpha) = -\tan^{-1}(M_{\alpha S}/M_{\alpha C})$$
$$= \eta_0 + \tan^{-1}\left[\tfrac{1}{2}\Gamma_\alpha/(E_{res} - E_\alpha)\right]. \tag{3.127}$$

When $\tilde{w}_0$ in Eq. (3.43) is replaced by its asymptotic form (Miller, 1970)

$$\tilde{w}_0 \sim w_0 \cos \eta_0 + w_1 \sin \eta_0, \tag{3.128}$$

Eq. (3.43) becomes

$$\Gamma_\alpha = 4 \cos^2 \eta_0 (M_{\alpha S} + M_{\alpha C} \tan \eta_0)^2. \tag{3.129}$$

Both Γ_α and η_0 are determined by the intersection of graphs of Γ_α versus η_0, assuming fixed values $M_{\alpha S}$ and $M_{\alpha C}$ in Eq. (3.129), for two or more stabilized states α. Figure 3.5 shows an example, given by Fels and Hazi (1971), in which three such curves intersect at a common point. When Γ_α and η_0 are known, E_{res} is determined by Eq. (3.127) for a single-channel resonance. Extension of this procedure to multichannel resonances is more difficult, because the Harris formula is limited to single-channel scattering, but it has been considered by Fels and Hazi (1972). For single-channel scattering this method requires evaluation of bound–free matrix elements, but only at a few energy values, and does not require free–free matrix elements.

Taylor and Hazi (1976) point out that if E_α is accurate, the phase shift at E_α is just $\eta_0 + \pi/2$. Then the value of Γ_α computed from Eq. (3.129) becomes an upper bound to the true width. Hence nonlinear parameters

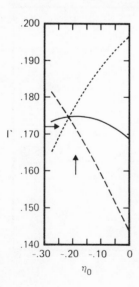

Figure 3.5. Dependence of Γ on η_0 for three different stabilized eigenfunctions (Fels and Hazi, 1971, Fig. 4).

in a trial wave function can be varied so as to minimize Γ_α computed in this way.

Hazi (1978) has shown that resonance widths can be deduced from calculations using quadratically integrable (L^2) functions only. For single-channel scattering, Eq. (3.43) defines a function of energy,

$$\Gamma_\alpha(E) = 4(\tilde{w}_0(E)|H - E_\alpha|\alpha)^2, \tag{3.130}$$

which takes on the value Γ_α at E_{res} corresponding to E_α. The functions $\tilde{w}_0(E)$ are required to be orthogonal to ϕ_α. As demonstrated in analysis of the stabilization method, an eigenfunction of H in an L^2 basis with matrix eigenvalue E_i approximates the continuum eigenfunction $w_0(E_i)$, except for a normalization factor, inside some completeness radius appropriate to the basis set. Since $(H - E_\alpha)\phi_\alpha$ is assumed to be adequately represented by this basis set, a similar expansion of $\tilde{w}_0(E)$ suffices to determine $\Gamma_\alpha(E)$, except for the normalization factor. Care must be taken in this argument that H is not diagonalized in a basis that describes both $\tilde{w}_0(E)$ and ϕ_α since Eq. (3.130) would vanish. This is avoided by use of projection operators (Hazi, 1978), as will be described below.

Hazi makes use of Stieltjes moment theory (Langhoff, 1974) to determine the function $\Gamma_\alpha(E)$ from the quantities

$$\gamma_{\alpha i} = 2\pi(u_i|H - E_\alpha|\alpha)^2, \tag{3.131}$$

computed for normalized functions u_i that are eigenfunctions, with energy eigenvalues ε_i, of the Hamiltonian H_{PP} in a space orthogonal to ϕ_α. The boundary conditions that define the continuum eigenfunctions $\tilde{w}_0(E)$ are such that these functions are normalized to give a uniform density of states with respect to energy. This property is used in the Green's function given by Eq. (3.35) (Fano, 1961). It follows from this that the cumulative width function

$$F_\alpha(E) = \int^E dE' \Gamma_\alpha(E') \tag{3.132}$$

can be approximated by the cumulative histogram sum

$$\tilde{F}_\alpha(E) = 0, \qquad 0 \geq E < \varepsilon_1, \tag{3.133}$$

$$\tilde{F}_\alpha(E) = \sum_{i=1}^{k} \gamma_{\alpha i}, \qquad \varepsilon_k < E < \varepsilon_{k+1},$$

defined at the rise points ε_k by

$$\tilde{F}_\alpha(\varepsilon_k) = \tfrac{1}{2}[\tilde{F}_\alpha(\varepsilon_{k^-}) + \tilde{F}(\varepsilon_{k^+})]$$

$$= \sum_{i=1}^{k-1} \gamma_{\alpha i} + \tfrac{1}{2}\gamma_{\alpha k}. \tag{3.134}$$

If a smooth curve is fitted to this histogram, its derivative is an approximation to the width function

$$\Gamma_\alpha(E) = dF_\alpha/dE \tag{3.135}$$

$$\cong d\tilde{F}_\alpha/dE. \tag{3.136}$$

Moment analysis (Langhoff and Corcoran, 1974; Hazi, 1978) can be used to replace the raw distribution

$$\{\gamma_{\alpha i}, \varepsilon_i; i = 1, \ldots, N\} \tag{3.137}$$

by a "principal representation" distribution

$$\{\gamma_{\alpha a}, \varepsilon_a; a = 1, \ldots, M\} \tag{3.138}$$

whose elements are uniquely determined by the first $2M$ inverse-power moments

$$\mu_k = \sum_{i=1}^{N} \gamma_{\alpha i}\varepsilon_i^{-k}, \tag{3.139}$$

where $M \ll N$. The moments converge to definite values as the original L^2 basis is extended, so the principal representation obtained using all converged moments contains all information conveyed by a calculation with a given basis.

In the method proposed by Hazi (1978), the function ϕ_α is obtained by a stabilization calculation or true variational calculation in a basis of functions orthogonal to any open-channel continuum with which ϕ_α interacts at energy E_α. The functions u_i are obtained by diagonalizing the matrix of the Hamiltonian in a basis of functions orthogonal to ϕ_α but otherwise unconstrained. A similar procedure is followed if several resonances of the same total symmetry are considered together.

Using this method, Hazi (1978) has computed the width of several resonances for electron scattering by He^+, He, and Mg. The results are in substantial agreement with data from other calculations using more conventional methods.

Hazi (1978) also gives a formula for an energy shift function, the principal value integral

$$\Delta_\alpha(E) = \frac{1}{2\pi} P \int \frac{\Gamma_\alpha(E')}{E - E'} \, dE', \tag{3.140}$$

which can be evaluated once $\Gamma_\alpha(E)$ is determined. The energy shift Δ_α is evaluated as $\Delta_\alpha(E_\alpha)$, or, more precisely, as $\Delta_\alpha(E_{\text{res}})$, where

$$E_{\text{res}} = E_\alpha + \Delta_\alpha(E_{\text{res}}). \tag{3.141}$$

Then the resonance width is $\Gamma_\alpha(E_{\text{res}})$.

3.5. Adiabatic Theory of Perturbed Target States

In many scattering processes, energy levels of the target system are split by a perturbation that is weak relative to the interaction responsible for the scattering. In the adiabatic approximation (Chase, 1956), an S matrix is computed without considering the perturbation, then it is transformed to a representation defined by the perturbed target states. This avoids the solution of close-coupling equations indexed by all of the perturbed states. In electron–atom scattering, this approximation can be applied to fine-structure excitation (Saraph, 1973) by transforming an S matrix obtained in LS coupling to a jj-coupling representation.

The adiabatic approximation neglects the effect on threshold scattering structures and on resonances of the energy-level splitting and energy shifts of the perturbed target states. A modified adiabatic theory that takes these effects into account has been proposed (Nesbet, 1979a) under the name "energy-modified" adiabatic theory, or EMA.

Since one of the most important applications of the adiabatic scattering theory is to electron-impact excitation of molecular rotational and vibrational states, the discussion here will use vibrational excitation of a diatomic molecule as an example. Electron–molecule rotational close-coupling calculations have been compared with the adiabatic approximation by Golden *et al.* (1971). Chandra and Temkin (1976) argue that the adiabatic approximation is not necessarily valid for vibrational excitation involving partial wave components of the electronic wave function that show resonance structure. The scattering delay time can be comparable to the period of vibrational motion. This violates the basic physical condition for validity of the adiabatic approximation, which requires the projectile velocity to be large compared with internal velocities of the target system.

For an electron scattered by a diatomic molecule, the Hamiltonian in body-fixed electronic coordinates $\mathbf{r}$, with internuclear distance R, is

$$H = T_n(R) + V(R) + T_e(\mathbf{r}) + V_{ne}(R, \mathbf{r}). \tag{3.142}$$

Here T_n is the nuclear kinetic energy operator, T_e is the electronic kinetic energy operator, $V(R)$ is the electronic energy of the target molecule (only one state is considered), and $V_{ne}(R, \mathbf{r})$ is the effective potential energy of the external electron. Only vibration will be treated here explicitly.

Defining the operator

$$H_n(R) = T_n(R) + V(R), \tag{3.143}$$

the eigenstates of nuclear motion are given by

$$H_n \chi_\mu(R) = E_\mu \chi_\mu(R). \tag{3.144}$$

At given total energy, a specific solution of the electronic and vibrational Schrödinger equation can be expanded in internal states as

$$\psi_\nu(R, \mathbf{r}) = \sum_\mu \chi_\mu(R)\psi_{\mu\nu}(\mathbf{r}). \tag{3.145}$$

The index ν is used to denote a particular incoming state, $\chi_\nu(R)$. The close-coupling equations for $\psi_{\mu\nu}(\mathbf{r})$ are

$$\sum_{\mu'} (\mu|T_e(\mathbf{r}) + V_{ne}(R, \mathbf{r}) - \varepsilon_\mu|\mu')\psi_{\mu'\nu}(\mathbf{r}) = 0, \qquad \text{all } \mu, \tag{3.146}$$

where the residual electronic energy in state μ is

$$\varepsilon_\mu = E - E_\mu. \tag{3.147}$$

In the fixed-nuclei approximation, electronic continuum wave functions are computed at each R, for arbitrary positive electronic energy ε, as solutions of

$$\{T_e(\mathbf{r}) + V_{ne}(R, \mathbf{r}) - \varepsilon\}\psi(\varepsilon; R, \mathbf{r}) = 0. \tag{3.148}$$

Since the nuclear coordinate R occurs only as a parameter, the functions

$$\psi_\nu(\varepsilon; R, \mathbf{r}) = \psi(\varepsilon; R, \mathbf{r})\chi_\nu(R) \tag{3.149}$$

satisfy the same equation.

The close-coupling amplitudes $\psi_{\mu\nu}(\mathbf{r})$ of Eq. (3.146) can be expressed formally in terms of the fixed-nuclei wave functions $\psi(\varepsilon; R, \mathbf{r})$. This is done by replacing the parameter ε, whose value in state μ is ε_μ, by the operator $E - H_n$. This substitution has no effect in Eq. (3.146), but replaces Eq. (3.148) by the formal operator equation

$$\{T_e(\mathbf{r}) + V_{ne}(R, \mathbf{r}) + H_n - E\}\psi(E - H_n; R, \mathbf{r}) = 0. \tag{3.150}$$

If the functions $\chi_\mu(R)$ form a complete set, this operator equation is equivalent to Eq. (3.146) if

$$\psi_{\mu\nu}(R) = (\mu|\psi(E - H_n; R, \mathbf{r})|\nu). \tag{3.151}$$

Equation (3.149) has been used here to define the specific solution ψ_ν.

In the limit of large r, the outgoing-wave part of ψ defines an S matrix, which can be symbolized by $S(\varepsilon; R)$. From Eq. (3.150), this becomes a unitary operator $S(E - H_n; R)$. Then Eq. (3.151) gives the S-matrix element connecting states χ_μ and χ_ν as

$$S_{\mu\nu} \cong (\mu|S(E - H_n; R)|\nu). \tag{3.152}$$

The usual adiabatic approximation (Chase, 1956) would estimate $S_{\mu\nu}$ in the form

$$S_{\mu\nu} \cong (\mu|S(\varepsilon; R)|\nu), \tag{3.153}$$

where ε is usually taken to be the energy of the incident particle. This approximation is expected to be valid only when the delay time is small compared with periods of internal motion of the target system. This requires

$$E - E_\mu \gg |E_\mu - E_\nu| \tag{3.154}$$

for energy levels involved in a transition but also can fail if there is a scattering resonance (Chandra and Temkin, 1976). When compared with Eq. (3.152), Eq. (3.153) clearly fails to take into account the internal motion.

Approximations based on Eq. (3.152) are considered in the EMA theory. If $S(\varepsilon; R)$ were independent of R, the operator $S(E - H_n; R)$ would be diagonal in the states χ_μ, as required by Born–Oppenheimer theory. If the dependence on R is weak, this implies for diagonal elements that

$$S_{\mu\mu} \cong (\mu|S(E - E_\mu; R)|\mu). \tag{3.155}$$

Since $S(\varepsilon; R)$ is a unit matrix for $\varepsilon = 0$, this formula has qualitatively correct threshold behavior: $S_{\mu\mu}$ approaches unity as $E \to E_\mu$. Unitarity requires that the transition elements $S_{\mu\nu}$ must approach zero at E_μ.

In the fixed-nuclei approximation, structural features in scattering cross sections occur as functions of the electronic energy ε in Eq. (3.148). At fixed R, structural features occur at the continuum threshold $\varepsilon(R) = 0$, or at a resonance energy

$$\varepsilon_{\text{res}}(R) = E_{\text{res}}(R) - V(R). \tag{3.156}$$

From Eqs. (3.147) and (3.152), the relevant residual energy $E - E_\mu$ or ε_μ is different in each internal state μ. Several important conclusions follow from this remark. In particular, threshold effects should occur at $\varepsilon_\mu = 0$ at each of the perturbed substate thresholds. Resonance structures should correspond to poles of the S matrix at complex energies given approximately, in the limit of weak coupling between internal states χ_μ, by mean values separately for each internal state,

$$W_\mu^{\text{res}} \cong (\mu|E_\mu + \varepsilon_{\text{res}}(R) - \tfrac{1}{2}i\Gamma(R)|\mu). \tag{3.157}$$

Here $\varepsilon_{\text{res}}(R)$ and the width $\Gamma(R)$ are computed in the fixed-nuclei approximation. This formula is equivalent to

$$W_\mu^{\text{res}} \cong (\mu|T_n(R) + E_{\text{res}}(R) - \tfrac{1}{2}i\Gamma(R)|\mu). \tag{3.158}$$

A single electronic resonance at each R becomes a *manifold* of resonances, one for each internal state μ, at complex energies averaged over the internal coordinate R and displaced by the nuclear kinetic energy $(\mu|T_n|\mu)$. Similar conclusions must hold for fine-structure perturbations in electron–atom scattering.

When the target states χ_μ cannot be considered to be uncoupled, a matrix approximation to Eq. (3.152) is required. While correct threshold energy dependence can be assured in diagonal elements by use of Eq. (3.155), it is more difficult to construct nondiagonal elements compatible with unitarity. A simple approximation with the correct properties can be expressed in terms of the K matrix, related to the S matrix by

$$S = (I + iK)(I - iK)^{-1}, \tag{3.159}$$

where I is a unit matrix. If K is real and symmetric then S is unitary.

A suitable approximation is the symmetric matrix

$$K_{\mu\nu} \cong (\mu|K(\varepsilon_{\mu\nu}; R)|\nu). \tag{3.160}$$

If $E_\mu > E_\nu$, $\varepsilon_{\mu\nu}$ is chosen so that $K_{\mu\nu}$ approaches zero correctly at the threshold $\varepsilon_\mu \to 0$. A simple choice is

$$\varepsilon_{\mu\nu} = (\varepsilon_\mu \varepsilon_\nu)^{1/2}, \tag{3.161}$$

which gives $K_{\mu\nu}$ correct threshold behavior for short-range potentials.

For an isolated fixed-nuclei electron scattering resonance, a more explicit representation can be used based on Eqs. (3.54) and (3.60). The background K matrix is

$$K_0 = \sin \eta_0 (\cos \eta_0)^{-1} \tag{3.162}$$

and the resonant K matrix is of the form

$$K_{\text{res}} = y(\tan \eta_{\text{res}}) y^{\dagger}, \tag{3.163}$$

where y is a real unit column vector whose components at fixed R are electronic partial wave amplitudes, and η_{res} is defined by

$$\tan \eta_{\text{res}}(R) = -\tfrac{1}{2}\Gamma(R)/(\varepsilon - \varepsilon_{\text{res}}(R)). \tag{3.164}$$

If K_0, y, Γ, and ε_{res} are obtained from analysis of $S(\varepsilon; R)$, they can be replaced by matrices, with $(\mu|K_0|\nu)$ given in symmetrized form by Eq. (3.160). The parameter ε must be replaced by $E - H_n$.

The resonance width Γ arises physically from barrier penetration and should be primarily a function of the residual energy ε rather than of the coordinate R. If Γ is parameterized as a function of ε, the resonance width at given R is $\Gamma(\varepsilon_{\text{res}}(R))$, from the residue of $\tan \eta_{\text{res}}$ at the resonance pole. The resulting symmetrized matrix expression for Eq. (3.164) is

$$(\mu|\tan \eta_{\text{res}}(R)|\nu) = -\tfrac{1}{2}\Gamma^{1/2}(\varepsilon_\mu)(\varepsilon - \varepsilon_{\text{res}})_{\mu\nu}^{-1}\Gamma^{1/2}(\varepsilon_\nu), \tag{3.165}$$

using the matrix inverse to

$$(\varepsilon - \varepsilon_{\text{res}})_{\mu\nu} = \varepsilon_\mu \delta_{\mu\nu} - (\mu|\varepsilon_{\text{res}}(R)|\nu). \tag{3.166}$$

Equation (3.165) has the correct structure for a resonance near a threshold. If all matrices were diagonal, Eq. (3.165) would agree with Eq. (3.157) for the complex energy corresponding to a pole of the perturbed S matrix.

The EMA approximation provides a rationale for correcting adiabatic electron–molecule cross sections near rotational excitation thresholds by modifying electron momenta in the cross-section formula, as proposed by Chang and Temkin (1970). A similar argument was used by LeDourneuf and Nesbet (1976) to compute fine-structure level collision strengths in electron scattering by atomic oxygen. The K matrix computed for elastic scattering in LS coupling was transformed to a jj-coupling representation (Saraph, 1972, 1973). Then excitation cross sections were evaluated as functions of energy values shifted by the fine-structure level splittings. This was done to ensure that the computed excitation cross sections would go to zero at the appropriate fine-structure thresholds.

Equation (3.155) indicates that when a degenerate target energy level is split by a weak perturbation, threshold structure should be repeated at each perturbed threshold. This provides a possible interpretation (Nesbet, 1977b) of structure observed in electron scattering by polar molecules. Excitation cross-section peaks occur at each vibrational threshold in e^-– HCl scattering (Rohr and Linder, 1975, 1976), and sharp downward steps are observed at these threshold energies in dissociative attachment (Ziesel et al., 1975; Abouaf and Teillet-Billy, 1977). It was suggested that the threshold excitation peaks were analogous to similar structure, enhanced by a virtual state, that occurs in e^-–He scattering, and that the dissociative attachment data showed Wigner threshold structures in the form of rounded steps of very small effective width.

Equation (3.165) has been tested in calculations of vibrational excitation structures in e^-–N_2 scattering (Nesbet, 1979a). The functions $\Gamma(\varepsilon)$ and $\varepsilon_{res}(R)$ were taken from model calculations by Birtwistle and Herzenberg (1971). The matrix $(\varepsilon - \varepsilon_{res})_{\mu\nu}$ was evaluated in a basis of the first 15 Morse oscillator eigenfunctions appropriate to the electronic ground state of N_2. Comparison with observed vibrational excitation structure (Ehrhardt and Willmann, 1967b) shows detailed agreement in peak spacing and in variations of the structure for different final states.

4

Computational Technique

Introduction

The variational formalism presented in Chapter 2 makes use of matrix elements of the operator $H - E$, where H is the $(N + 1)$-electron Hamiltonian, between antisymmetrized electronic wave functions. Such matrix elements are required in Eq. (2.9) for the matrix variational method. Similar matrix elements occur in Eq. (2.251), for the $(N + 1)$-electron hybrid method based on the Schwinger variational principle.

In this chapter, the formalism needed to evaluate these matrix elements will be described. Particular reference is made to Lyons *et al.* (1973), who give computational details of an implementation of the matrix variational method. The description given here is modified to make use of LS eigenfunctions.

4.1. Reduction to One- and Two-Electron Integrals

The $(N + 1)$-electron Hamiltonian is of the form

$$H = \sum_i h(i) + \sum_{ij} v(ij), \tag{4.1}$$

where $i = 1, \ldots, N + 1$ and $i < j \le N + 1$. Any $(N + 1)$-electron wave function can be expressed as a linear combination of Slater determinants, defined as normalized antisymmetrized products of one-electron orbital functions. Matrix elements of H in a basis of Slater determinants are given by simple formulas (Condon and Shortley, 1935; Nesbet, 1965).

In electron–atom scattering theory, it is convenient to construct $(N + 1)$-electron wave functions in terms of N-electron parent states. The

119

classification given in Eq. (1.7) for $(N + 1)$-electron basis functions for the bound or Hilbert space component Ψ_Q of the scattering wave function makes use of an N-electron reference Slater determinant Φ_0. The occupied orbital functions of Φ_0 are the N orthonormal functions $\{\phi_i\}$, considered to be functions of all variables, space and spin, that describe a single electron. A complementary orthonormal set of orbitals $\{\phi_a\}$, unoccupied in Φ_0, is assumed.

States of the N-electron target atom correspond to wave functions that can be represented as linear combinations of N-electron Slater determinants

$$\Phi_{ij\ldots}^{ab\ldots}, \qquad i < j < \cdots < N < a < b < \cdots, \qquad (4.2)$$

also defined with respect to Φ_0. The N-electron Hamiltonian is given by Eq. (4.1), with $i = 1, \ldots, N$ and $i < j \le n$. Formulas will be given first for matrix elements of $H(N)$, the N-electron Hamiltonian. The modifications necessary for evaluation of the three types of $(N + 1)$-electron matrix elements (bound–bound, bound–free, and free–free) will then be presented in a common formalism.

The general rules for matrix elements of the N-electron Hamiltonian in a basis of Slater determinants can be derived most easily by using the fact that a normalized determinant Φ can be expressed as

$$\Phi = \det \Pi = \mathscr{A}\Pi, \qquad (4.3)$$

where Π is a product of orbital functions, each referring to a different electron. The operator $\mathscr{A}$ is the total antisymmetrizing operator

$$\mathscr{A} = (N!)^{-1/2}\{1 - \Sigma P_{ij} + \Sigma P_{ijk} - \cdots\} \qquad (4.4)$$

summed over all permutations of the electron indices. Since

$$\mathscr{A}^2 = (N!)^{1/2}\mathscr{A}, \qquad (4.5)$$

this defines a projection operator. Since the Hamiltonian is symmetric in the electron indices, it commutes with $\mathscr{A}$. Hence (Löwdin, 1955)

$$(\Phi_\mu|H|\Phi_\nu) = (\Pi_\mu|H|(N!)^{1/2}\mathscr{A}\Pi_\nu). \qquad (4.6)$$

Matrix elements between Slater determinants defined by Eq. (4.2) and the reference determinant Φ_0 follow from Eq. (4.6) and the orthonormality of the orbitals. They are

$$(\Phi_{ijk}^{abc}|H|\Phi_0) = 0, \qquad \text{more than two substitutions,} \qquad (4.7)$$

$$(\Phi_{ij}^{ab}|H|\Phi_0) = (ab|v(1 - P_{12})|ij) = (ab|\bar{v}|ij), \qquad (4.8)$$

$$(\Phi_i^a|H|\Phi_0) = (a|h|i) + \sum_j (aj|\bar{v}|ij) = (a|\mathscr{H}_0|i), \qquad (4.9)$$

$$(\Phi_0|H|\Phi_0) = \sum_i (i|h|i) + \sum_{ij} (ij|\bar{v}|ij)$$

$$= \frac{1}{2} \sum_i [(i|h|i) + (i|\mathscr{H}_0|i)]. \tag{4.10}$$

Summations over indices $ij \dots$ here always denote the range $i < j < \cdots \le N$. The definition of the auxiliary operator $\bar{v}$ is clear from Eq. (4.8). The operator $\mathscr{H}_0$ in Eq. (4.9) is the Hartree–Fock operator for the reference state Φ_0,

$$\mathscr{H}_0 = h + \sum_{i(occ)} (i|\bar{v}|i). \tag{4.11}$$

All matrix elements in a determinantal basis can be deduced from these formulas and expressed in terms of $\bar{v}$ and $\mathscr{H}_0$. For arbitrary Slater determinants, matrix elements $H_{\nu\mu}$ are expressed at first in terms of the operator

$$\mathscr{H}_\mu = h + \sum_{i(\mu)} (i_\mu|\bar{v}|i_\mu), \tag{4.12}$$

where the indices refer to the orbitals occupied in Φ_μ. Because $\mathscr{H}_\mu - \mathscr{H}_0$ is a sum of operators like $(i|\bar{v}|i)$, the result can be rewritten in terms of $\mathscr{H}_0$ and $\bar{v}$.

A useful formal result that follows from this analysis is that diagonal energy differences between Slater determinants can be expressed entirely in terms of one- and two-electron matrix elements that involve only the orbitals occupied in one determinant but not in the other. If diagonal matrix elements of H are denoted by

$$D_{ij\dots}^{ab\dots} = (\Phi_{ij\dots}^{ab\dots}|H|\Phi_{ij\dots}^{ab\dots}) \tag{4.13}$$

then

$$D_{ij\dots}^{ab\dots} - D_0 = (a|\mathscr{H}_0|a) + (b|\mathscr{H}_0|b) + \cdots - (i|\mathscr{H}_0|i) - (j|\mathscr{H}_0|j) - \cdots$$
$$+ (ab|\bar{v}|ab) + \cdots + (ij|\bar{v}|ij) + \cdots - (ai|\bar{v}|ai) - \cdots. \tag{4.14}$$

Thus the diagonal energy difference is given by the obvious difference of one-electron energies, corrected by adding all interactions among the new orbitals $(ab|\bar{v}|ab)$, by adding all interactions among the old orbitals $(ij|\bar{v}|ij)$, and by subtracting all mixed pair interactions $(ai|\bar{v}|ai)$. Equation (4.14) remains valid as a difference between diagonal energies of $H(N + 1)$ and $H(N)$. In the notation of Eq. (1.7), Φ^a is defined by adding ϕ_a to the occupied set of Φ_0, and Φ_i is defined by removing orbital ϕ_i. The energy differences given by Eq. (4.14) are

$$D^a - D_0 = (a|\mathscr{H}_0|a), \tag{4.15}$$

$$D_i - D_0 = -(i|\mathscr{H}_0|i). \tag{4.16}$$

These formulas were given originally by Koopmans (1933). Familiar examples of Eq. (4.14) are

$$D_i^a - D_0 = (a|\mathcal{H}_0|a) - (i|\mathcal{H}_0|i) - (ai|\bar{v}|ai) \tag{4.17}$$

$$D_{ij}^{ab} - D_0 = (a|\mathcal{H}_0|a) + (b|\mathcal{H}_0|b) - (i|\mathcal{H}_0|i) - (j|\mathcal{H}_0|j) + (ab|\bar{v}|ab) + (ij|\bar{v}|ij)$$

$$- (ai|\bar{v}|ai) - (aj|\bar{v}|aj) - (bi|\bar{v}|bi) - (bj|\bar{v}|bj). \tag{4.18}$$

For general matrix elements $H_{\mu\nu}$ in which Slater determinants Φ_μ and Φ_ν are both defined as in Eq. (1.7) or (4.2) by virtual excitation from a reference function Φ_0, it is necessary to express Φ_μ as a virtual excitation of Φ_ν before using Eqs. (4.7)–(4.9) and (4.14). In these formulas, $\mathcal{H}_0$ is to be replaced by $\mathcal{H}_\nu$. The matrix element $H_{\mu\nu}$ is analyzed by a simple algorithm that uses a substitution notation equivalent to Eq. (4.2). An N-electron Slater determinant is specified by a list of $\bar{n}$ virtual excitations a/i of Φ_0. This can be encoded as a list of $2\bar{n} + 1$ integers

$$\bar{n}, a_1, i_1, a_2, i_2, \ldots, a_{\bar{n}}, i_{\bar{n}}, \tag{4.19}$$

with the convention $a_1 < a_2 < \cdots$ and $i_1 < i_2 < \cdots$.

Given two such lists,

$$\Phi_\mu: m, a_{1\mu}, i_{1\mu}, \ldots, a_m, i_m,$$
$$\Phi_\nu: n, a_{1\nu}, i_{1\nu}, \ldots, a_n, i_n, \tag{4.20}$$

the algorithm for $H_{\mu\nu}$ produces a similar list defining Φ_μ as a virtual excitation of Φ_ν. A phase factor $(-1)^p$ is defined by the number of interchanges of orbital indices required to match as many as possible of the occupied orbitals of Φ_μ to those of Φ_ν.

The code for Φ_ν is scanned, and each element i_ν is compared with all elements i_μ of the Φ_μ code. If a match is found, the element i_μ is replaced by a_ν corresponding to the matched element i_ν. If i_ν differs from all i_μ, the pair of elements i_ν, a_ν are added in the stated order to the end of the Φ_μ list, and m is increased by one.

When the Φ_ν code is exhausted, the resulting modified Φ_μ list is scanned and each element $i_{j\mu}$ (in the extended list) is compared with every element $a_{b\mu}$. If a match is found, element $a_{j\mu}$ is replaced by $a_{b\mu}$, elements in positions $i_{j\mu}$ and $a_{j\mu}$ are deleted, the sign of $(-1)^p$ is reversed ($p = 0$ initially), and m is decreased by one.

It can easily be verified that the final modified Φ_μ list is the virtual excitation code for Φ_μ relative to Φ_ν. The degree of virtual excitation is given by the final value of m, and the phase factor for matrix elements is the final value of $(-1)^p$. Matrix elements of H are given by Eq. (4.8) if final $m = 2$ and by Eq. (4.9) with $\mathcal{H}_0$ replaced by $\mathcal{H}_\nu$ if final $m = 1$. Other nondiagonal matrix elements vanish. Diagonal elements are evaluated relative to the reference state energy D_0 by Eq. (4.14).

An $(N + 1)$-electron Slater determinant, used as a basis function for component Ψ_Q of scattering wave function, can be defined by Eq. (1.7). This is encoded as a list of $2\bar{n} + 1$ integers, with $\bar{n} > 0$,

$$\bar{n}, a_1, i_1, \ldots, a_{\bar{n}}, -1 \tag{4.21}$$

as in Eq. (4.19). Although Eq. (4.21) defines an $(N + 1)$-electron function in terms of the N-electron reference function Φ_0, the algorithm described above works correctly for $(N + 1)$-electron matrix elements, if the orbital indices i, a are positive numbers (Lyons $et\ al.$, 1973).

Bound–free matrix elements of $H(N + 1)$, given by Eq. (2.11), are of the form

$$M_{\mu,ip} = \sum_\sigma (\Phi_\mu|H - E|\mathscr{A}\Phi_\sigma\phi_{ip})c_{\sigma p}, \tag{4.22}$$

where

$$\Theta_p = \sum_\sigma \Phi_\sigma c_{\sigma p} \tag{4.23}$$

is a target-state wave function expanded in a basis of N-electron Slater determinants $\{\Phi_\sigma\}$. The radial factor of the continuum orbital function ϕ_{ip} is $r^{-1}F_{ip}(r)$, defined by its asymptotic form as in Eq. (2.1) and by orthogonality of $F_{ip}(r)$ to all radial functions with $l = l_p$ in the assumed orthonormal bound basis. Since ϕ_{ip} is orthogonal to all bound orbitals, the matrix elements combined in Eq. (4.22) can be evaluated by the algorithm described above for matrix elements between Slater determinants.

Since all of the matrix elements combined in Eq. (4.22) refer to the same open-channel orbitals ϕ_{ip} $(i = 0, 1)$, it is convenient to analyze them in terms of the $(N + 1)$-electron Slater determinant Φ_μ and the N-electron determinant Φ_σ. These are encoded by integer lists as in Eqs. (4.21) and (4.20), respectively, expressing both in terms of virtual excitations of Φ_0. The antisymmetrized function $\mathscr{A}\Phi_\sigma\phi_{ip}$ could be encoded by appending (ip), -1 to the code for Φ_σ. It can easily be verified that the algorithm for matrix elements gives correct formulas for bound–free matrix elements if the Φ_σ code is extended in this way.

The same results are obtained more simply by applying the algorithm directly to the codes for Φ_μ and Φ_σ, except that the place of (ip) is taken in the relative excitation code by "-1". The level of relative excitation Δn of Φ_μ with respect to Φ_σ, using this convention, is always at least one. Because of orbital orthogonality it follows, as in Eqs. (4.8) and (4.9), that the only nonzero matrix elements are of the form

$$\pm(b_1|\mathscr{H}_\sigma|F), \qquad \Delta n = 1, \tag{4.24}$$

$$\pm(b_1 b_2|\bar{v}|b_3 F), \qquad \Delta n = 2, \tag{4.25}$$

where b_a denotes a bound orbital and F denotes F_{ip} ($i = 0, 1$). It should be noted that $\mathscr{H}_\mu$ could be used instead of $\mathscr{H}_\sigma$ in Eq. (4.24) because

$$(bb|\bar{v}|bF) \equiv 0, \tag{4.26}$$

from the definition of $\bar{v}$.

Free–free matrix elements, given by Eq. (2.12), are of the form

$$M_{ij}^{pq} = \sum_\sigma \sum_\tau c_{\sigma p}(\mathscr{A}\Phi_\sigma\phi_{ip}|H - E|\mathscr{A}\Phi_\tau\phi_{jq})c_\tau. \tag{4.27}$$

Because the open-channel orbitals are not quadratically integrable, matrix elements M_{ij}^{pq} are not defined unless

$$\tfrac{1}{2}k_q^2 = E - E_q = \varepsilon_q, \tag{4.28}$$

where E_q is an eigenvalue of the target configuration interaction matrix of $H(N)$ in the basis $\{\Phi_\sigma\}$,

$$(\Theta_p|H(N)|\Theta_q) = E_q\delta_{pq}. \tag{4.29}$$

Equation (4.28) can be used to express Eq. (4.27) as

$$M_{ij}^{pq} = (\Theta_p|H - E_q|\Theta_q)(\phi_{ip}|\phi_{jq}) + [\mathscr{A}\Theta_p\phi_{ip}|H - \varepsilon_q|\mathscr{A}\Theta_q\phi_{jq}]. \tag{4.30}$$

Here the square-bracket notation is used to denote a matrix element evaluated *as if* $(\phi_{ip}|\phi_{jq})$ were zero for all values of the indices. Terms in $(\phi_{ip}|\phi_{jq})$ are explicitly collected together in the first term of Eq. (4.30). Equation (4.29) implies that this first term vanishes regardless of the definition of $(\phi_{ip}|\phi_{jq})$. A more precise argument can be based on truncating all integrals at some large value of the radial coordinate r, then passing to the limit $r \to \infty$. The result is

$$M_{ij}^{pq} = [\mathscr{A}\Theta_p\phi_{ip}|H - \varepsilon_q|\mathscr{A}\Theta_q\phi_{jq}]$$

$$= \sum_\sigma \sum_\tau c_{\sigma p}[\mathscr{A}\Phi_\sigma\phi_{ip}|H - \varepsilon_q|\mathscr{A}\Phi_\tau\phi_{jq}]c_{\tau q}. \tag{4.31}$$

The matrix elements in Eq. (4.31) are obtained by applying the algorithm described above to the encoded integer lists for Φ_σ and Φ_τ, then appending (ip), (jq) to the code describing the relative virtual excitation of Φ_σ with respect to Φ_τ. If this relative excitation index is Δn, the only nonzero matrix elements are of the form

$$\pm(\phi_{ip}|\mathscr{H}_\tau - \varepsilon_q|\phi_{jq}), \qquad \Delta n = 0, \tag{4.32}$$

$$\pm(b_1\phi_{ip}|\bar{v}|b_2\phi_{jq}), \qquad \Delta n = 1, \tag{4.33}$$

where b_a denotes a bound orbital.

4.2. Vector Coupling and Angular Integrals

The nonrelativistic electronic Hamiltonian H commutes separately with the total orbital angular momentum operator $\mathbf{L}$ and with the total spin operator $\mathbf{S}$. At given total energy E, the degenerate set of eigenfunctions can be transformed into an equivalent set of $LS\pi$ eigenfunctions. Matrix elements of H in a basis of such functions vanish unless all angular and parity quantum numbers of the initial and final functions are equal. The non-vanishing matrix elements are independent of the values of M_L and M_S, the components of $\mathbf{L}$ and $\mathbf{S}$, respectively, along a space-fixed coordinate axis.

In electron scattering by light atoms, these properties decouple the scattering amplitudes for different $(N + 1)$-electron $LS\pi$ states from each other. This appears in the general formula for a scattering cross section, Eq. (1.29), which is a sum of terms evaluated independently for states with different $LS\pi$ values.

In matrix variational calculations, the $(N + 1)$-electron basis functions used for Ψ_Q of Eq. (1.50) can be LS-coupled functions expressed as linear combinations of Slater determinants. Calculations can be carried out independently for each set of $LS\pi$ quantum numbers. Because all matrix elements of the Hamiltonian are independent of M_L and M_S, only the particular values

$$M_L = L, \qquad M_S = S \tag{4.34}$$

need be considered. As indicated in Section 1.5 above, the structure of Ψ_Q can be expressed in terms of electronic configurations defined by virtual excitations of a reference configuration, usually that of the target atom ground state. Methods are needed to construct LS eigenfunctions as linear combinations of the Slater determinant basis functions for a specified configuration and to evaluate matrix elements of H in the basis of LS eigenfunctions.

Efficient and general methods can be based on the formal theory of angular momentum and irreducible tensors (Rose, 1957; Edmonds, 1957; Brink and Satchler, 1968). The construction of angular momentum eigenfunctions in atomic physics is usually based on progressive vector coupling, building up N-electron configurations one electron at a time. The Racah algebraic technique, as described by Fano (1965), has been widely used in electron scattering calculations. A computer program for configuration interaction calculations, published by Hibbert (1975), implements this method.

An alternative method (Nesbet, 1961) constructs a complete orthonormal set of LS eigenfunctions for a given configuration, without reference to the sequence of parent states required in progressive vector coupling. This eliminates the need for auxiliary tables or computer subroutines for fractional parentage coefficients or angular momentum recoupling coefficients.

For a general angular momentum operator **J**, the basic equation of this method is

$$J_+ \psi_{jj} \equiv 0, \tag{4.35}$$

where

$$J_+ = J_x + iJ_y. \tag{4.36}$$

Equation (4.35) is a special case of the "ladder" equation

$$J_+ \psi_{jm} = \hbar[(j - m)(j + m + 1)]^{1/2} \psi_{j,m+1}, \tag{4.37}$$

valid for all angular momentum operators and their eigenfunctions ψ_{jm}. The eigenvalue of $\mathbf{J}^2$ is $\hbar^2 j(j + 1)$ and the eigenvalue of J_z is $\hbar m$, where $-j \le m \le j$. Here j and m are either both integers or both integers plus $1/2$.

To construct LS eigenfunctions, Eq. (4.35) is solved in the matrix representation defined by a given configuration. Matrix elements of L_+ connect basis functions (M_L, M_S) with functions $(M_L + 1, M_S)$. Matrix elements of S_+ connect (M_L, M_S) with $(M_L, M_S + 1)$. These matrix elements are obtained directly from matrix elements of the one-electron operators l_+ and s_+ given by Eq. (4.37) for the appropriate quantum numbers. The resulting system of homogeneous linear equations is solved for the coefficients that determine a set of orthonormal LS eigenfunctions for the particular case given by Eq. (4.34). Subsequent evaluation of matrix elements of general N-electron or $(N + 1)$-electron operators is simplified if the solution vectors are obtained in a particular canonical form.

The algebraic problem is to obtain vectors x_μ as solutions of the homogeneous system

$$\sum_{i=1}^{n} m_{ji} x_{\mu i} = 0, \qquad j = 1, \ldots, n_E, \tag{4.38}$$

in the canonical form defined by

$$x_{\mu i} = \delta_{\mu i}, \qquad i \le \mu, \tag{4.39}$$

and subject to orthogonality

$$\sum_{i=1}^{n} x_{\mu i} x_{\nu i} = d_\mu \delta_{\mu \nu}. \tag{4.40}$$

The dimension of the orthonormal basis is n, the number of distinct Slater determinants in the given configuration whose M_L and M_S values satisfy Eq.

Table 4.1. Example: Construction of LS Eigenfunctions

		1	2	3	4	5	6	
				i				
				Equations				
		0	0	0	0	1	1	
		0	$(2/3)^{1/2}$	0	1	1	0	
		1	0	1	1	0	0	
		1	$(2/3)^{1/2}$	0	0	0	0	
		1	0	0	−1	0	1	
				Pivots				
	6	2	0	1	0	0	1^a	
$\lambda =$	5	−2	0	−1	0	1^a		
	4	1	0	1	1^a			
	2	$(3/2)^{1/2}$	1^a					
				Solutions y				
$\mu =$	3			1^a	−1	1	−1	
	1	1^a	$-(3/2)^{1/2}$	0	−1	2	−2	
				Coefficients a/x				d_μ
$\mu =$	3	5/4	0	1^a	−1	1	−1	4
	1	1^a	$-(3/2)^{1/2}$	−5/4	1/4	3/4	−3/4	21/4

^{a}Marks first or last nonzero element of a row vector.

(4.34) for given L and S. Since parity is a property of a configuration, all functions considered have the same parity.

If n and n_E are small numbers, Eq. (4.38) can be solved trivially by Gauss elimination. In practice, these dimensions can be several hundred or more, and the matrix m_{ji} can be quite large. Since this matrix is relatively sparse (many zero elements) and tends to become more so as Gauss elimination proceeds, a sparse matrix method can be used (Nesbet, 1978c). Methods for inhomogeneous systems are not immediately useful because the homogeneous system is necessarily singular. The number of solutions n_S is known only when $n_P = n - n_S$ independent pivot equations have been obtained at the conclusion of Gauss elimination.

The method used is illustrated by the simple example of ^2D functions from the configuration d^3, shown in Table 4.1. The arithmetic is that of Gauss elimination, reducing the given matrix of linear equations to a matrix of pivot vectors. The pivot vectors are required to be of the form

$$\pi_{\lambda i} = \delta_{\lambda i}, \qquad \lambda \le i, \tag{4.41}$$

and also to satisfy

$$\pi_{\lambda \lambda'} = 0, \qquad \lambda \ne \lambda'. \tag{4.42}$$

Each successive original equation vector is reduced to this form or eliminated completely. All elements $m_{j\lambda}$ for any previously defined pivot index λ are reduced to zero by subtracting multiples of pivot vectors π_λ, in decreasing order of λ. If any nonzero element m_{ji} remains, the vector is divided by the nonzero element of largest index i to become a new pivot vector $\pi_{\lambda'}$ with $\lambda' = i$, satisfying Eq. (4.41). Nonzero coefficients $\pi_{\lambda\lambda'}$ in previous pivot vectors with $\lambda > \lambda'$ are then eliminated by subtracting an appropriate multiple of $\pi_{\lambda'}$ so that Eq. (4.42) is satisfied.

Since computer memory space may not be available for the $n_E \times n$ elements of m_{ji}, equation vectors $\mathbf{m}_j$ can be read one at a time from auxiliary memory. The array of pivot vector π_λ can be stored in condensed form, with zeroes eliminated. Nonzero elements are defined by a list of column indices. The packed lists of column indices and nonzero elements can be stored in a workspace of maximum available length, indexed by a vector of pointers $P(\lambda)$, to locate each pivot list. The pointer element is set to zero when π_λ is not defined. Updated pivot vectors can be added at the open end of the workspace, which needs to be repacked only when exhausted.

Table 4.1 shows a final block of n_P pivot vectors π_λ, in unpacked format. Even in this simple case, if index integers require half the space of real numbers, the storage space required for the packed pivot vectors is less than that required for the full array of $n_P = 4$ pivot vectors, each of length $n = 6$.

The pivot matrix determines solution vectors $\mathbf{y}_\mu$ of the reduced equations

$$\sum_{i=1}^{n} \pi_{\lambda i} y_{\mu i} = 0, \qquad \text{all } \lambda, \tag{4.43}$$

for n_P values of λ and $n_S = n - n_P$ values of μ. From the structure of π_λ, linearly independent solutions $\mathbf{y}_\mu$ are obtained by requiring

$$y_{\mu i} = \delta_{\mu i}, \qquad i \le \mu, \tag{4.44}$$

where μ is a column index not included in the set of pivot indices $\{\lambda\}$. The condition

$$y_{\mu\mu'} = 0, \qquad \mu' \ne \mu, \tag{4.45}$$

can also be required, where μ' is also not in the set $\{\lambda\}$. With these conditions, the nonzero elements of $\mathbf{y}_\mu$ are

$$y_{\mu\mu} = 1,$$
$$y_{\mu\lambda} = -\pi_{\lambda\mu}, \qquad \eta > \mu. \tag{4.46}$$

This determines the row-vectors $\mathbf{y}_\mu$ directly from the nonzero column vectors of the array of pivot vectors. This can be verified in the example shown in Table 4.1. Since $\{y_{\mu i}\}$ is not sparse, condensation is not useful, so the full space $n_S \times n$ is required for it. In practice this is not restrictive since $n_S \ll n_E$.

Final processing of $\{y_{\mu i}\}$ produces a table of coefficients a/x in the canonical form required by this method (Nesbet, 1961). Vectors $\mathbf{x}_\mu$ are obtained from $\mathbf{y}_\mu$ by Schmidt orthogonalization, retaining Eq. (4.44) but not Eq. (4.45). The adjoint vectors $\mathbf{a}_\mu$ satisfy

$$\sum_{i=1}^{n} a_{\mu i} x_{\mu' i} = \delta_{\mu\mu'}, \tag{4.47}$$

where

$$a_{\mu i} = \delta_{\mu i}, \qquad i \geq \mu, \tag{4.48}$$

and

$$a_{\mu i} = 0, \qquad i \text{ not in } \{\mu\}. \tag{4.49}$$

The vectors $\mathbf{a}_\mu$ are obtained by back substitution, equivalent to inverting a triangular matrix. An example of this canonical coefficient table is shown in Table 4.1. Because of Eqs. (4.44) and (4.48), $\{a_{\mu i}\}$ and $\{x_{\mu i}\}$ can be stored in a single rectangular array. Normalization constants d_μ are defined by

$$d_\mu = \sum_{i=\mu}^{n} x_{\mu i}^2. \tag{4.50}$$

The a/x coefficients are used to evaluate matrix elements of any rotationally invariant operator H in the form (Nesbet, 1961)

$$(A\mu|H|B\nu) = \left(\frac{d_\nu}{d_\mu}\right)^{1/2} \sum_i \sum_j x_{\mu i}^A a_{\nu j}^B (Ai|H|Bj). \tag{4.51}$$

Here A and B index configurations, and i and j index basis functions, all with common values of M_L, M_S and parity. Indices μ, ν refer to orthonormal LS eigenfunctions defined by the coefficient tables $(a/x)_A$ and $(a/x)_B$. The summation indices in Eq. (4.51) are limited to the ranges $\mu \leq i \leq n_A$ and $1 \leq j \leq \nu$, with j in the set of solution indices $\{\nu\}$ for configuration B. Since $n_S \ll n$ in general, the restricted range of index j, which is a consequence of the underlying projection-operator formalism (Löwdin, 1955), significantly reduces the number of matrix elements $(Ai|H|Bj)$ to be considered.

When Eq. (4.51) is combined with the analysis given in Section 4.1, it provides formulas for matrix elements of H between LS-coupled functions as linear combinations of one- and two-electron integrals. Since the algorithm given here constructs functions with the special property indicated in Eq. (4.34), the target-atom wave functions Θ_p will have this property, as will the basis functions for the Hilbert-space component Ψ_Q of the $(N + 1)$-electron scattering wavefunction. Some special consideration is needed for the coupled functions $\mathscr{A}\Theta_p\phi_{ip}$ in Eq. (2.4) in order to take advantage of the special structure of the directly computed target functions Θ_p.

The LS-coupled function denoted by $\mathscr{A}\Theta_p\phi_{ip}$ is of the general form

$$\mathscr{A}\Theta_p\phi_{ip} = (d_p)^{-1/2}\left\{[\mathscr{A}\Theta_p\phi_{ip}]_0 + \sum_{\gamma \neq 0} x_\gamma[\mathscr{A}\Theta_p\phi_{ip}]_\gamma\right\}, \qquad (4.52)$$

where

$$d_p^2 = 1 + \sum_{\gamma \neq 0} x_\gamma^2. \qquad (4.53)$$

Here $[\mathscr{A}\Theta_p\phi_{ip}]$ is an antisymmetrized product of the standard LS eigenfunction Θ_p, for which $M_L' = L_p$ and $M_S' = S_p$, and an orbital function ϕ_{ip}, with orbital quantum numbers l_p and

$$\begin{aligned} m_l &= L - L_p, \\ m_s &= S - S_p. \end{aligned} \qquad (4.54)$$

The total $(N + 1)$-electron quantum numbers are $LS\pi$, and the target-state quantum numbers are $L_pS_p\pi_p$. The remaining terms in Eq. (4.52) correspond to all other possible values of m_l, m_s and M_L', M_S' for which

$$\begin{aligned} M_L' + m_l &= M_L = L, \\ M_S' + m_s &= M_S = S. \end{aligned} \qquad (4.55)$$

The coefficients x_γ are defined so that $\mathscr{A}\Theta_p\phi_{ip}$ is proportional to the projection (Löwdin, 1955) of $[\mathscr{A}\Theta_p\phi_{ip}]_0$ onto the vector space of LS eigenfunctions with quantum numbers $LS\pi$. It can be shown (Nesbet, 1961) that this projected function is given by Eq. (4.52) multiplied by an additional factor of $(d_p)^{-1/2}$. Because only two angular momenta are coupled, separately for orbital angular momenta (L', l_p) and for spin $(S', 1/2)$, only one vector-coupling scheme is possible. This implies that $\mathscr{A}\Theta_p\phi_{ip}$ is identical with the usual vector-coupled function obtained from Θ_p and ϕ_{ip}, and the coefficients in Eq. (4.52) can be evaluated as standard vector-coupling coefficients. In particular, if the coefficient for coupling angular momenta j_1 and j_2 to resultant j is denoted by

$$(j_1m_1j_2m_2|jm), \qquad (4.56)$$

then the coefficient of the leading term in Eq. (4.52) is

$$d_p^{-1/2} = (L_pL_pl_pL - L_p|LL)(S_pS_p\tfrac{1}{2}S - S_p|SS). \qquad (4.57)$$

Explicit formulas for these vector-coupling coefficients (Edmonds, 1957, p. 43) give

$$d_p = \frac{(L_p + l_p + L + 1)!\,(L_p - l + L)!\,(S_p + \frac{1}{2} + S + 1)!\,(S_p - \frac{1}{2} + S)!}{(2L_p)!\,(2L + 1)!\,(2S_p)!\,(2S + 1)!}. \qquad (4.58)$$

This can be computed in the form

$$d_p = \frac{2L_p + 1}{2L + 1}\left[\prod_{k=0}^{L_p + l_p - L}\left(\frac{2L + 1 + k}{2L_p + 1 - k}\right)\right]\frac{2S_p + 1}{2S + 1}\left[\prod_{k=0}^{S_p + \frac{1}{2} - S}\left(\frac{2S + 1 + k}{2S_p + 1 - k}\right)\right].$$
$$(4.59)$$

Projection operator analysis (Nesbet, 1961) shows that bound–free matrix elements defined by Eq. (2.11) can be evaluated for LS eigenfunctions Φ_μ without requiring values of any of the coefficients in Eq. (4.52) other than that of the leading term, given by Eq. (4.57). The required formula is

$$(\Phi_\mu|H - E|\mathcal{A}\Theta_p\phi_{ip}) = \left(\frac{d_p}{d_\mu}\right)^{1/2}\sum_j x_{\mu j}(\Phi_i|H - E|[\mathcal{A}\Theta_p\phi_{ip}]_0), \qquad (4.60)$$

where the coefficients $x_{\mu j}$ are taken from the canonical a/x table for the expansion of Φ_μ as a linear combination of Slater determinants $\{\Phi_j\}$ from a given configuration, and d_μ is defined by Eq. (4.50). Equation (4.60) is a special case of Eq. (4.51) since Eq. (4.52) defines an LS eigenfunction as the normalized projection of $[\mathcal{A}\Theta_p\phi_{ip}]_0$. The integrals in the sum indicated in Eq. (4.60) are reduced to one- and two-electron integrals by the analysis given in Section 4.1, leading to Eqs. (4.24) and (4.25) for specific matrix elements.

Free–free matrix elements between LS eigenfunctions can be evaluated by an application of the Wigner–Eckart theorem (Wigner, 1927; Eckart, 1930; Brink and Satchler, 1968, p. 56). As indicated in Eq. (4.52), the vector-coupled function $\mathcal{A}\Theta_p\phi_{ip}$ is a linear combination of functions with specified values of L_p, S_p, l_p, and s_p but with all values of M'_L, M'_S, m_l, and m_s that are compatible with total quantum numbers $M_L = L$, $M_S = S$. The function Θ_p and its single-configuration component functions Φ_σ are constructed in a standard state with $M'_L = L_p$ and $M'_S = S_p$. The matrix elements of $H - E$ between functions described by

$$M'_L = L_p, \qquad \Lambda = M_L = L_p + m_{lp},$$
$$(4.61)$$
$$M'_S = S_p, \qquad \Sigma = M_S = S_p + m_{sp},$$

can be constructed directly by the analysis given in Section 4.1. The Wigner–Eckart theorem is used to express matrix elements between LS

eigenfunctions with $L \geq \Lambda$ and $S \geq \Sigma$ as linear combinations of those computed for functions defined by Eqs. (4.61).

In order to derive the transformation formula, it is convenient to consider the simplified case of orbital angular momentum only. Free–free matrix elements can be symbolized by

$$(L_p M_p l_p m_p | H - E | L_q M_q l_q m_q), \tag{4.62}$$

in a basis of antisymmetrized product functions that are not vector-coupled. The directly computed matrix elements, corresponding to Eqs. (4.61), can be denoted by

$$(I_{pq})_\Lambda = (L_p L_p l_p \Lambda - L_p | H - E | L_q L_q l_q \Lambda - L_q). \tag{4.63}$$

The vector-coupling coefficients of Eq. (4.56) define an orthogonal transformation between uncoupled functions $(j_1 m_1 j_2 m_2)$ and coupled functions (jm). If they are used to express the uncoupled basis functions of Eq. (4.63) in terms of coupled functions $(L\Lambda)$, this gives

$$(I_{pq})_\Lambda = \sum_{L \geq \Lambda} (L_p L_p l_p \Lambda - L_p | L\Lambda)(I_{pq})_L (L\Lambda | L_q L_q l_q \Lambda - L_q). \tag{4.64}$$

Here the matrix element between vector-coupled functions $(L\Lambda)$ is denoted by

$$(I_{pq})_L = (p; L\Lambda | H - E | q; L\Lambda), \qquad L \geq \Lambda. \tag{4.65}$$

This is independent of Λ because of the assumed spherical symmetry of H. Equation (4.64) is an example of the Wigner–Eckart theorem. It shows that the required integrals $(I_{pq})_L$ serve as a set of reduced matrix elements (Edmonds, 1957, p. 75; Brink and Satchler, 1968, p. 56) for all of the matrix elements of Eq. (4.62).

Equation (4.64) defines a linear transformation between elements of the vectors $(I_{pq})_L$ and $(I_{pq})_\Lambda$. The transformation matrix is triangular, with dimension defined by the usual angular momentum conditions,

$$\max(|L_p - l_p|, |L_q - l_q|) \leq L \leq \min(L_p + l_p, L_q + l_q). \tag{4.66}$$

Linearly independent values of $(I_{pq})_\Lambda$ are obtained only for Λ in the same range as L. With this condition, Eq. (4.64) defines a nonsingular transformation matrix that can be inverted to give $(I_{pq})_L$ as a linear combination of elements $(I_{pq})_\Lambda$.

When spin is taken into account, the generalization of Eq. (4.64) is of the form

$$(I_{pq})_{\Lambda\Sigma} = \sum_{L \geq \Lambda} \sum_{S \geq \Sigma} \Gamma_{\Lambda\Sigma,LS}(I_{pq})_{LS}, \tag{4.67}$$

where

$$\Gamma_{\Lambda\Sigma,LS} = \gamma_{\Lambda L}\gamma_{\Sigma S}, \tag{4.68}$$

$$\gamma_{\Lambda L} = (L_p L_p l_p \Lambda - L_p | L\Lambda)(L\Lambda | L_q L_q l_q \Lambda - L_q), \tag{4.69}$$

$$\gamma_{\Sigma S} = (S_p S_p \tfrac{1}{2}\Sigma - S_p | S\Sigma)(S\Sigma | S_q S_q \tfrac{1}{2}\Sigma - S_q). \tag{4.70}$$

The required vector-coupling coefficients can be evaluated from standard formulas [e.g., Rose, 1957, Eq. (3.18)].

Because both matrices $\gamma_{\Lambda L}$ and $\gamma_{\Sigma S}$ are triangular, the matrix Γ is also triangular. Hence Eq. (4.67) can be used directly to compute elements $(I_{pq})_{LS}$ from given values of $(I_{pq})_{\Lambda\Sigma}$ by back substitution. The formal expression is

$$(I_{pq})_{LS} = \sum_{\Lambda \le L} \sum_{\Sigma \le S} (\Gamma^{-1})_{LS,\Lambda\Sigma}(I_{pq})_{\Lambda\Sigma}. \tag{4.71}$$

The integrals in $(I_{pq})_{\Lambda\Sigma}$ are reduced to one- and two-electron integrals by the analysis leading to Eqs. (4.32) and (4.33) for specific matrix elements. In Eq. (4.71), the range of Λ and L is given by Eq. (4.66). The corresponding limits on Σ and S are

$$\max(|S_p - \tfrac{1}{2}|, |S_q - \tfrac{1}{2}|) \le \Sigma, S \le \min(S_p + \tfrac{1}{2}, S_q + \tfrac{1}{2}). \tag{4.72}$$

The one- and two-electron integrals obtained by the analysis given above are expressed in terms of orbital functions that consist of a radial factor, a spin factor, and an orbital angular momentum factor in the form of a spherical harmonic function $Y_{lm}(\theta, \phi)$. All integrals reduce to linear combinations of integrals over the operators $h(i)$ and $v(ij)$ of Eq. (4.1). For scattering by an atom, the one-electron operator $h(i)$ is spherically symmetrical, so one-electron integrals reduce to radial integrals, multiplied by the factors

$$(lm_l m_s | l'm_l'm_s') = \delta(ll')\delta(m_l m_l')\delta(m_s m_s'). \tag{4.73}$$

The two-electron interaction is the interelectronic Coulomb potential, which can be expanded in the well-known series

$$v(12) = \frac{1}{r_{12}} = \sum_\lambda \frac{4\pi}{2\lambda + 1} \frac{r_<^\lambda}{r_>^{\lambda+1}} \sum_{\mu=-\lambda}^{+\lambda} Y_{\lambda\mu}^*(\theta_1, \phi_1) Y_{\lambda\mu}(\theta_2, \phi_2). \tag{4.74}$$

Here $r_>$, $r_<$ denote, respectively, the greater or lesser of r_1 and r_2. Angular integration gives

$$(ab|v|cd) = \sum_\lambda c^\lambda(a;c)c^\lambda(d;b)[ac|db]^\lambda \delta(m_{sa}m_{sc})\delta(m_{sd}m_{sb}), \tag{4.75}$$

where $[ac|db]^\lambda$ is a reduced matrix element of $v(12)$, the generalized

Slater parameter

$$[ac|db]^\lambda = \int_0^\infty dr_1 \int_0^\infty dr_2 \frac{r_<^{\lambda+2}}{r_>^{\lambda-1}} R_a^*(r_1)R_c(r_1)R_d(r_2)R_c^*(r_2). \quad (4.76)$$

The radial factor of ϕ_a is denoted by $R_a(r)$ here. In Eq. (4.75), the angular integrals are the Gaunt coefficients (Condon and Shortley, 1935, pp. 178 and 179)

$$c^\lambda(lm; l'm') = \left(\frac{4\pi}{2\lambda+1}\right)^{1/2}$$

$$\times \int_0^{2\pi} d\phi \int_0^\pi \sin\theta\, d\theta Y_{lm}^*(\theta, \phi) Y_{\lambda, m-m'}(\theta, \phi) Y_{l'm'}(\theta, \phi). \quad (4.77)$$

This integral can be expressed in the form (Edmonds, 1957, pp. 46 and 63)

$$c^\lambda(lm; l'm') = (-1)^m[(2l+1)(2l'+1)]^{1/2}\begin{pmatrix} l & l' & \lambda \\ 0 & 0 & 0 \end{pmatrix}\begin{pmatrix} l & l' & \lambda \\ -m & m' & m-m' \end{pmatrix}, \quad (4.78)$$

in terms of the $3j$ symbols of Wigner or as

$$c^\lambda(lm; l'm') = (-1)^{m'}\frac{[(2l+1)(2l'+1)]^{1/2}}{2\lambda+1}$$

$$\times (l\ \ 0\ \ l'\ \ 0|\lambda\ \ 0)(l, -m, l', m'|\lambda, m-m') \quad (4.79)$$

in terms of vector-coupling coefficients. It follows from symmetry properties of the $3j$ symbols that $l + l' + \lambda$ must be even, and

$$c^\lambda(l'm'; lm) = (-1)^{m'-m}c^\lambda(lm; l'm') \quad (4.80)$$

$$= c^\lambda(l', -m'; l, -m). \quad (4.81)$$

Because Gaunt coefficients $c^\lambda(a; c)$ are required for every two-electron integral and because they have a limited range of arguments governed by the orbital l values, in practical calculations it is convenient to compute all required Gaunt coefficients prior to the main calculation and to tabulate them for later reference. Recurrence relations can easily be derived that simplify their evaluation. Because of Eqs. (4.80) and (4.81), the only coefficients required are those with

$$l \geq l', \qquad m \geq 0, \qquad -l' \leq m' \leq l'. \quad (4.82)$$

In the (ll') subtable, there are $l' + 1$ entries $c^\lambda(lm; l'm')$ for each (mm'), corresponding to

$$\lambda = l - l', l - l' + 2, \ldots, l + l'. \qquad (4.83)$$

4.3. Radial Integrals

Variational calculations of electron–atom scattering require two different kinds of radial factors for orbital wave functions. Bound orbitals can be represented as linear combinations of functions of the form $r^{-1}\eta_a(r)$ where

$$\eta_a(r) = N_a r^{n_a} \exp(-\zeta_a r). \qquad (4.84)$$

Open-channel orbitals require radial functions that are oscillatory at large r. The simplest functions of this kind are the spherical Bessel functions

$$j_l(kr) = (\pi/2kr)^{1/2} J_{l+1/2}(kr), \qquad (4.85)$$

which are solutions of the free-particle Schrödinger equation for orbital angular momentum l. Functions normalized according to Eqs. (2.1) for their asymptotic forms are given by

$$S(r) = k^{1/2} r j_l(kr) \sim k^{-1/2} \sin(kr - \tfrac{1}{2}l\pi), \qquad (4.86)$$

$$C(r) = -k^{1/2} r y_l(kr) \sim k^{-1/2} \cos(kr - \tfrac{1}{2}l\pi). \qquad (4.87)$$

Here $y_l(kr)$ is the corresponding irregular function (Abramowitz and Stegun, 1964, Section 10.1). In practical calculations, continuum basis functions must be used that are regular at the origin. One method is to multiply $y_l(kr)$ by a factor that vanishes sufficiently rapidly for small r but approaches unity at large r. Another method (Armstead, 1968) is to use regular spherical Bessel functions of higher order, in the form

$$C(r) = k^{1/2} r \left[j_{l+1}(kr) + \frac{l+1}{kr} j_{l+2}(kr) \right]. \qquad (4.88)$$

The asymptotic form agrees with Eq. (4.87) up to terms in r^{-2}.

One- and two-electron radial matrix elements required for variational calculations can be evaluated for bound orbitals of exponential form, as in Eq. (4.84). Efficient procedures can be based on simple recurrence formulas (Nesbet, 1963). Methods for computing radial integrals that involve spherical Bessel functions have been considered by Harris and Michels (1969b, 1971), by Lyons and Nesbet (1969, 1973) using Armstead's form for $C(r)$. Evaluation of integrals containing functions $C(r)$ defined by applying a regularizing factor to $y_l(kr)$ has been considered by Oberoi et al. (1972). Computer programs for the radial integrals, with Armstead's $C(r)$, have been published by Smith and Truhlar (1973) and by Abdallah and Truhlar (1975).

For electron–ion scattering, Coulomb wave functions must be used as continuum basis functions. The spherical Bessel functions are limiting forms of Coulomb–Bessel functions. Because of this, much of the analytical theory of the radial integrals can be generalized. Formulas for integrals involving Coulomb–Bessel functions have been derived by Bottcher (1970) and by Ramaker (1972).

Hybrid methods, considered in Section 2.6, may provide a practical alternative to the direct use of continuum basis functions for Coulomb or electric dipole scattering. In these methods, an exact solution of the coupled asymptotic equations is combined with variational approximation of the inner part of the scattering wave function. The exact asymptotic solution imposes a natural cutoff on radial integrals involving the continuum basis functions. Given a finite cutoff radius r_0, it may be computationally efficient to evaluate the residual integrals by direct numerical integration.

Unless a hybrid method is used, radial integrals must be evaluated over the full range of r. For oscillatory integrands, this is very inefficient compared with the analytic methods that have been developed, especially for spherical Bessel functions as continuum basis functions. The analysis of Lyons and Nesbet (1969, 1973), incorporated into the programs of Smith and Truhlar (1973) and of Abdallah and Truhlar (1975), will be outlined here. Only the case of spherical Bessel functions is considered.

Radial integrals that occur in a variational calculation are of the types

$$\text{1-electron bound–bound:} \quad (a|h - \varepsilon|b), \tag{4.89a}$$

$$\text{1-electron bound–free:} \quad (a|h - \varepsilon|F_p), \tag{4.89b}$$

$$\text{1-electron free–free:} \quad (F_p|h - \varepsilon|F_q), \tag{4.89c}$$

$$\text{2-electron bound–bound:} \quad [ab|cd]^\lambda, \tag{4.89d}$$

$$\text{2-electron bound–free:} \quad [ab|cF_p]^\lambda, \tag{4.89e}$$

$$\text{2-electron free–free Coulomb:} \quad [ab|F_pF_q]^\lambda, \tag{4.89f}$$

$$\text{2-electron free–free exchange:} \quad [aF_p|bF_q]^\lambda. \tag{4.89g}$$

Here h denotes the radial one-electron Hamiltonian operator. The two-electron integrals are defined by Eq. (4.76). Indices a, etc., refer to the bound radial functions of Eq. (4.84). The symbol F_p denotes a continuum basis function $S(r)$ or $C(r)$ for channel p. The bound–bound integrals can be evaluated accurately and efficiently in closed form (e.g., Nesbet, 1963). The one-electron bound–free and free–free integrals are simplified by the fact that the functions F are asymptotically proportional to continuum eigenfunctions of the one-electron Hamiltonian. The two-electron bound–free integrals are the most numerous of those involving continuum functions, but they can be expressed explicitly as Gaussian hypergeometric functions (Lyons and Nesbet, 1969; Bottcher, 1970; Ramaker, 1972). This makes it possible to use many results of classical analysis to simplify their accurate evaluation. The two-electron free–free integrals are much less numerous. The Coulomb integrals can be reduced to forms related to the bound–free integrals. The exchange integrals present special difficulties and require separate analysis (Lyons and Nesbet, 1973; Ramaker, 1972).

Lyons and Nesbet (1969) give formulas for reduction of the continuum integrals of Eqs. (4.89) to six basic radial integrals, defined as follows:

$$G(\lambda; p|k, \alpha) = \int_0^\infty j_\lambda(kr) r^{p-\lambda} e^{-\alpha r} dr, \tag{4.90a}$$

$$H(\lambda, \mu; p|k_1, k_2, \alpha) = \int_0^\infty j_\lambda(k_1 r) j_\mu(k_2 r)^{p-\lambda-\mu} e^{-\alpha r} dr, \tag{4.90b}$$

$$I(\lambda, \mu; p|k_1, k_2) = \int_0^\infty j_\lambda(k_1 r) j_\mu(k_2 r) r^{p-\lambda-\mu} dr, \tag{4.90c}$$

$$V(\lambda, \mu; p, q|k_1, k_2, \alpha) = \int_0^\infty j_\lambda(k_1 r) j_\mu(k_2 r) A_p(\alpha, r) r^{q-\lambda-\mu} dr, \tag{4.90d}$$

$$W(\lambda; p, q|k, \alpha, \beta) = \int_0^\infty j_\lambda(kr) A_p(\alpha, r) r^{q-\lambda} e^{-\beta r} dr, \tag{4.90e}$$

$$X(\lambda, \mu; p, q|k_1, k_2, \alpha, \beta) = \int_0^\infty dr_1 j_\lambda(k_1 r_1) r_1^{p-\lambda} e^{-\alpha r_1}$$

$$\times \int_{r_1}^\infty dr_2 j_\mu(k_2 r_2) r_2^{q-\mu} e^{-\beta r_2}. \tag{4.90f}$$

In these equations, λ, p, and q are nonnegative integers. The parameters k_1 and k_2 are nonnegative real numbers. The parameters α and β may be complex, but their real parts are positive.

The auxiliary function $A_p(\alpha, r)$ is defined for complex α by

$$A_p(\alpha, r) = \int_r^\infty s^p e^{-\alpha s} ds, \tag{4.91}$$

if Re $(\alpha > 0)$. Integration by parts gives

$$A_{p+1}(\alpha, r) = \frac{1}{\alpha} r^{p+1} e^{-\alpha r} + \frac{p+1}{\alpha} A_p(\alpha, r). \tag{4.92}$$

Together with the starting value

$$A_0(\alpha, r) = \frac{1}{\alpha} e^{-\alpha r}, \tag{4.93}$$

this gives recurrence formulas for the integrals W and V of Eqs. (4.90) in terms of the integrals G and H, respectively.

The integrals I are a limiting case of the H integrals. A special formula is required for $I(\lambda, \mu; p|k, k)$, when $k_1 = k_2$, and is given by the Weber–Schafheitlin integral in a simple closed form [Lyons and Nesbet, 1969, Eq. (59)]. The H integrals can be computed by one of two different procedures, both using G integrals with complex argument α (Lyons and Nesbet, 1969). If $p \leq 2\lambda$ and $p \leq 2\mu$ and otherwise if α in the definition of H is greater than k_1 and k_2, a recurrence procedure must be used, starting from G integrals of complex argument. The recurrence formulas are obtained from the formula, valid for complex z,

$$(2l + 1)j_l(z) = z j_{l-1}(z) + z j_{l+1}(z). \tag{4.94}$$

This is used for both j_λ and j_μ in the H integral starting from

$$j_0(z) = z^{-1} \sin z = z^{-1} \operatorname{Im}(e^{iz}),$$
$$j_{-1}(z) = z^{-1} \cos z = z^{-1} \operatorname{Re}(e^{iz}). \tag{4.95}$$

If p is greater than 2λ or 2μ and α is less than k_1 and k_2, H can be expressed as a finite sum of G integrals of complex argument α.

Except for the exchange integrals X, all of the others reduce to forms requiring evaluation of G integrals, in general for complex argument α. The procedure given by Lyons and Nesbet (1969) for the G integrals is valid for all values of their arguments. The relationship to hypergeometric functions shows that the analytic character of these integrals is different in two different ranges of $p/2\lambda$. If $p > 2\lambda$ they are rational functions of k^2/α^2; if $p \leq 2\lambda$ they are transcendental functions. This implies that recurrence relations cannot be carried from the rational to the transcendental region. The coefficient of one term in the Gauss recurrence relations for hypergeometric functions, when applied to the G integrals, goes to zero on the boundary $p = 2\lambda$.

Because the recurrence relation reduces to two terms on this boundary, a single value of G on the boundary suffices to begin recurrence into the rational region. The required starting values are

$$G(\lambda; 2\lambda + 1|k, \alpha) = \frac{k^\lambda (2\lambda)!!}{(\alpha^2 + k^2)^{\lambda+1}}, \tag{4.96}$$

$$G(\lambda; 2\lambda + 2|k, \alpha) = \frac{k^\lambda \alpha (2\lambda + 2)!!}{(\alpha^2 + k^2)^{\lambda+2}}. \tag{4.97}$$

The double factorial notion here denotes a product of successive even or odd integers up to the indicated limit. The recurrence formula to be used for decreasing values of λ and fixed p in the rational region is given by Lyons and Nesbet [1969, Eq. (44)].

In the transcendental region, $p \leq 2\lambda$, two recurrence procedures are required to avoid loss of numerical accuracy. If $|\alpha| < k$, a three-term recurrence formula connecting $(\lambda - 1, p - 1)$, (λ, p), and $(\lambda + 1, p + 1)$ can be used, starting from

$$G(0; 0|k, \alpha) = \frac{1}{k} \tan^{-1}\left(\frac{k}{\alpha}\right), \tag{4.98}$$

$$\lim_{p \to 0} k_p G(\lambda - 1; p - 1|k, \alpha) = \frac{k^\lambda}{(2\lambda - 1)!!}. \tag{4.99}$$

All G integrals for $p < \lambda$ are obtained from the set $G(\lambda; \lambda|k, \alpha)$ by other three-term recurrence formulas. Integrals for $p > \lambda$ are given by the reflection formula [Lyons and Nesbet, 1969, Eq. (49)]

$$G(\lambda; p|k, \alpha) = \frac{p!}{(2\lambda - p)!} (\alpha^2 + k^2)^{\lambda - p} G(\lambda; 2\lambda - p|k, \alpha). \tag{4.100}$$

If $|\alpha| \geq k$, starting values for the two largest values of λ required are obtained from the hypergeometric series, an infinite series with argument $k^2/(\alpha^2 + k^2)$. Then the set of integrals $G(\lambda; \lambda|k, \alpha)$ is filled in by downward recurrence on λ, and the remaining integrals are computed as for $|\alpha| < k$.

The exchange integrals X of Eq. (4.90f) are especially difficult to evaluate accurately and efficiently. The integrand is oscillatory in two dimensions, which precludes direct numerical quadrature. Recurrence formulas can introduce rapid loss of numerical accuracy, and series expansions have regions of convergence that are difficult to determine as functions of all eight parameters. Lyons and Nesbet (1973) developed a practical procedure for these integrals based on the use of six different techniques in different ranges of the parameters.

The only simple formula for the X integrals is for the important special case

$$X(\lambda, \lambda; p, p|k_1, k_1, \alpha, \alpha) = \tfrac{1}{2}[G(\lambda; p|k_1, \alpha)]^2 \tag{4.101}$$

Two different infinite series can be derived, each useful in a different range of parameters. When $p > 2\lambda$ or $q > 2\mu$, the X integral can be expressed as a finite sum of G integrals of complex argument α. It was shown by Harris and Michels (1969b), for special values of the parameters, that the X integral can be transformed into a contour integral in the complex plane such that a nonoscillatory integrand is obtained, so that numerical quadrature is feasible. A generalization of this result gives a numerical quadrature formula that requires values of W integrals of complex argument. A quite complicated recurrence procedure completes the repertory of methods. An algorithm for selecting the appropriate method for specified values of the parameters, based on extensive numerical tests, is given by Lyons and Nesbet (1973). The proposed procedure is incorporated with minor modifications in the published program of Abdallah and Truhlar (1975).

In the analysis of matrix elements of the $(N + 1)$-electron Hamiltonian, individual integrals occur that appear to be undefined or infinite. They are one-electron matrix elements of the nuclear Coulomb attraction $-Z/r$ and two-electron Coulomb integrals with $\lambda = 0$, as given by Eq. (4.89f). Detailed analysis (Lyons and Nesbet, 1969) shows that there is exact cancellation between these terms when the full wave function for electron scattering by a neutral atom is taken into account. The physical basis for this result is that the nuclear Coulomb potential is completely screened at large r by the electronic charge distribution. As a practical procedure, one-electron bound–free and free–free integrals over $-Z/r$ can be set to zero if the corresponding terms in the two-electron Coulomb integrals are omitted.

4.4. Structure of a Computer Program

The various stages of calculation required in applications of the matrix variational method can most easily be described with reference to the structure of a computer program. Harris and Michels (1971) have outlined the steps necessary for a full calculation. A particular program package, which has been used for many of the calculations to be described below, uses computational and data-handling algorithms described by Lyons et al. (1973). A master diagram of the computational procedure, modified to use LS eigenfunctions, is shown in Fig. 4.1.

The program CON/360 serves as a monitor and driver for all other programs. It also provides various data-handling services through subroutine calls. CON/360 manages a general scratch data filing system. It also controls dynamical overlay so that program segments can be loaded, entered, or deleted on request from the currently operating program.

The initial program ATOMS, and several others not indicated in Fig. 4.1, read or modify input data. The initial states of an electron–atom scattering calculation make use of algorithms and programs designed for bound-state calculations (Nesbet, 1963). A matrix Hartree–Fock calculation is carried out for a specified target atom state by program SCF, using integrals of one- and two-electron terms in the atomic Hamiltonian operator that are evaluated over a basis of exponential radial functions, Eq. (4.84), by program ONECI. This calculation defines not only the occupied orbitals of the target

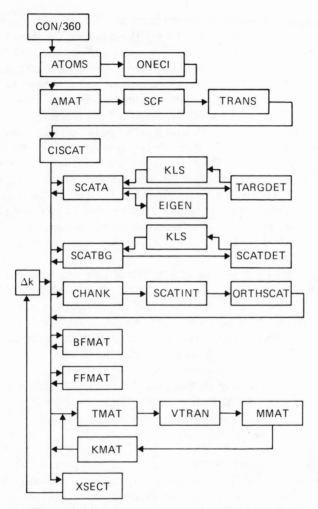

Figure 4.1. Map of programs for matrix variational method.

atom reference state, but also a set of orthonormal "unoccupied" orbitals, used later in the calculations to represent virtual excitations of the target atom state. The program TRANS transforms reduced matrix elements of the two-electron Coulomb potential into the basis of orthonormal orbitals defined by SCF.

The program CISCAT is the driving program for electron–atom scattering calculations.

States of the target atom are specified by supplying a list of configurations, defined in terms of virtual excitations of a target reference configuration. The target reference state quantum numbers are $L' = M'_L$ and $S' = M'_S$, where M'_L and M'_S are the quantum numbers of a Slater determinant Φ_0, specified by SCF. A maximum channel orbital angular quantum number $\bar{l}$ and a range of values of the total quantum numbers L, S, and π, the latter being ± 1, are also specified. This information suffices to define allowed ranges of the quantum numbers of N-electron target atom states to be constructed from the specified configurations.

The program SCATA builds up a list of possible target state quantum numbers (L', S', π'). The programs driven by SCATA (TARGDET,KLS,EIGEN) diagonalize the matrix of the N-electron Hamiltonian over all states of the specified target configurations with these quantum numbers. Only states with $M'_L = L'$ and $M'_S = S'$ are considered, using the analysis given in Section 4.2. The energy eigenvalues E_p are sorted by increasing value and the eigenvectors are saved for later reference. The program KLS is similar to KDB, as described by Lyons et al. (1973), but is modified for the use of LS-coupled functions.

The programs driven by SCATBG (SCATDET and KLS) construct the bound–bound matrix $H_{\mu\nu}$ in a basis of $(N + 1)$-electron LS-coupled functions, constructed as described in Section 4.2 from a master list of $(N + 1)$-electron configurations. The configuration list is generated as described in Sections 1.5 and 1.6. All values of total quantum numbers $LS\pi$ in a specified range are considered, and the matrix $H_{\mu\nu}$ is sectioned according to these quantum numbers.

Within an outer loop that defines a sequence of reference channel k values, program CHANK examines the target state energy eigenvalues obtained by SCATA and determines a list of open channels at total energy

$$E = E_0 + \tfrac{1}{2}k^2. \qquad (4.102)$$

Here E_0 is the lowest-energy eigenvalue for target states with the same quantum numbers $(L'S'\pi')$ as the reference state, and k is the current reference channel k value. A condensed list of open-channel quantum numbers and k_p values is constructed, together with sublists of virtual excitation codes, LS-coupling tables, and target state eigenvectors required to specify the states Θ_p for the open channels.

Channel orbital integrals for the continuum basis functions F_{ip} are computed by SCATINT using methods described in Section 4.3. These raw integrals are transformed by ORTHSCAT to a basis of orthonormal bound orbitals and orthogonalized continuum orbitals.

The bound–free and free–free matrices are constructed, respectively, by BFMAT and FFMAT. These programs use the methods described in Sections 4.1 and 4.2.

The matrix m_{ij}^{pq}, defined by Eq. (2.9), is constructed separately for each set of total quantum numbers $(LS\pi)$ and is used to compute the corresponding K matrix. This is indicated by the loop in Fig. 4.1 that includes programs TMAT, VTRAN, MMAT, and KMAT.

Because the bound–bound matrix may be quite large, computing the inverse matrix, as required for direct evaluation of Eq. (2.9), may be impractical. For this reason, m_{ij}^{pq} is constructed indirectly by means of a triangular factorization of the bound–bound matrix for given total quantum numbers,

$$M_{\mu\nu} = H_{\mu\nu} - E\delta_{\mu\nu}. \tag{4.103}$$

An algorithm is used (Nesbet, 1971) that adapts the well-known Cholesky factorization to the Hermitian but not positive definite matrix $M_{\mu\nu}$. A lower triangular matrix $T_{\mu\alpha}$ is defined by

$$M_{\mu\nu} = \sum_{\alpha} T_{\mu\alpha}\sigma_{\alpha}T_{\alpha\nu}^{\dagger}, \tag{4.104}$$

where σ_{α} is a diagonal matrix with elements ± 1. All matrix elements here are real. For a real matrix $T_{\nu\alpha}$ the adjoint $T_{\alpha\nu}^{\dagger}$ is just the transposed matrix. The matrix σ_{α} is represented by a list of integer index pointers to the relatively small number of elements $\sigma_{\alpha} = -1$. This number, n_{σ}, is the number of negative eigenvalues of $M_{\mu\nu}$. This property can be used in a resonance search procedure (Nesbet and Lyons, 1971). When $M_{\mu\nu}$ can be contained in main computer memory, it is replaced element by element by $T_{\mu\nu}$. Otherwise, main memory is used as a buffer area, and $T_{\mu\nu}$ is built up in sequential segments. Program TMAT constructs $T_{\mu\nu}$ using this algorithm.

Program VTRAN constructs an auxiliary rectangular matrix $B_{\alpha,ip}$ such that

$$\sum_{\sigma} T_{\mu\alpha}B_{\alpha,ip} = M_{\mu,ip} \tag{4.105}$$

or

$$B_{\alpha,ip} = \sum_{\mu} (T^{-1})_{\alpha\mu}M_{\mu,ip}, \tag{4.106}$$

where $M_{\mu,ip}$ is the bound–free matrix. Because T is triangular, B can be constructed by sequential processing of both $T_{\mu\alpha}$ and $M_{\mu,ip}$. In general, both of these matrices can be too large for storage in main computer memory. When $M_{\mu,ip}$ can be contained in main memory, it is replaced element by element by $B_{\alpha,ip}$. Otherwise, main memory is used as a buffer area and $B_{\alpha,ip}$ is built up in sequential segments.

In terms of the auxiliary matrix B, Eq. (2.9) reduces to

$$m_{ij}^{pq} = M_{ij}^{pq} - \sum_{\alpha} B_{ip,\alpha}^{\dagger} \sigma_{\alpha} B_{\alpha,jq} \tag{4.107}$$

$$= M_{ij}^{pq} - \sum_{\alpha} \sigma_{\alpha} B_{\alpha,ip} B_{\alpha,jq}. \tag{4.108}$$

This sum is evaluated by program MMAT in a single sequential scan of the matrix $B_{\alpha,ip}$. The relatively small free–free matrix M_{ij}^{pq} is stored in main memory and converted to m_{ij}^{pq} in place.

Program KMAT computes the K matrix from m_{ij}^{pq}, using one or more of the variational methods described in Section 2.4. The K matrix is diagonalized, and the eigenphases and eigenchannel vectors are stored for each set of total quantum numbers $(LS\pi)$. Program XSECT uses these data to compute scattering cross sections.

5

Applications to One-Electron Atoms

Introduction

Hydrogen is the unique atom for which bound-state wave functions are known exactly. The alkali metal atoms, with a single valence electron, can also be conveniently included in the class of "one-electron" atoms. A model that represents the closed-shell positive ion core of these atoms by an effective static potential is well justified, so that low-energy electron scattering is essentially a two-electron problem for both hydrogen and the alkali metals.

The $1s$ ground state of hydrogen has the special property that the lowest excited p state requires a change of principal quantum number . The $(2s)^2 S$ excited state of hydrogen is more closely analogous to the ground states of alkali metal atoms since the odd parity state $(2p)^2 P^0$ has the same energy. For an alkali metal atom, the ground-state $(ns)^2 S$ interacts strongly with the low-lying state $(np)^2 P^0$ through an electric dipole matrix element. This produces a very large polarizability and a strong polarization potential. The exact $2s$, $2p$ degeneracy in excited hydrogen (neglecting the Lamb shift) represents a limiting case described by a nonvanishing static electric dipole moment rather than by infinite polarizability.

Variational calculations of $e^- - H$ scattering have been carried out over a broad energy range that extends well beyond the ionization threshold (13.6 eV). Corresponding calculations of electron scattering by alkali metal atoms have been limited primarily to elastic scattering. The electron–hydrogen calculations have been reviewed in detail by Callaway (1978b).

145

5.1. Hydrogen: Elastic Scattering

The correct treatment of the electric dipole polarizability α_d of the target atom is essential to an accurate calculation of low-energy electron scattering. The partial wave Born formula, Eq. (3.102), indicates that for $l > 0$ the leading term in the expansion of the phase shift $\eta_l(k)$ for small k is proportional to α_d. Hence a fractional error in α_d affects the higher-order phase-shifts proportionately. If α_d is neglected, as it is in the static exchange approximation, high-order phase shifts are grossly underestimated at low energies. Detailed analysis shows that the resulting error in the differential cross section is most serious at small scattering angles.

The accurate variational calculations of Schwartz (1961b), which represented a direct solution of the continuum two-electron problem for open-channel s-waves, necessarily contained an accurate treatment of the polarization potential. These calculations used two-electron basis functions expressed in terms of r_1, r_2, and r_{12}, as in the bound-state calculations of Hylleraas (1928). Because Schwartz used the Kohn variational method without modification, values of phase shifts had to be deduced from the directly computed values exemplified by Figs. 2.1 and 2.2. These figures indicate the practical difficulties inherent in this method due to spurious singularities and slow convergence, especially at low scattering energies. Nevertheless, by smoothing out the anomalous singularities and by extrapolating the data to apparent convergence, Schwartz obtained the values of phase shifts listed in Table 5.1 for ^1S and ^3S scattering states. The figures in parentheses give the estimated uncertainty in the last-quoted digit.

Similar calculations were carried out by Armstead (1968) for the $^1P^0$ and $^3P^0$ scattering states. The inferred phase shifts, with error indications, are included in Table 5.1. Similar calculations were carried out for p- and d-wave phase shifts by Gailitis (1965b).

Absolute measurements of the e^-–H differential cross section in the elastic scattering range 0.5 to 8.7 eV were made by Williams (1975). The experimental accuracy was given as $\pm 6\%$. From Eq. (1.34), the differential cross section for elastic scattering is given for ground-state hydrogen by

$$\frac{d\sigma}{d\Omega} = \sum_{S=0}^{1} \frac{2S + 1}{4} |f_S|^2, \tag{5.1}$$

where

$$f_S(k, \theta) = \frac{1}{k} \sum_{L=0}^{\infty} (2L + 1) \exp(i\eta_{LS}) \sin \eta_{LS} P_L(\cos \theta). \tag{5.2}$$

Here η_{LS} denotes the phase shift in a scattering state with total quantum

Table 5.1. Phase Shifts η_{LS} for Electron–Hydrogen Elastic Scattering (Radians)[a]

$k(a_0^{-1})$	^1S	^3S	$^1P^0$	$^3P^0$	^1D	^3D	$^1F^0$	$^3F^0$
				State				
0.1	2.5530 (10)	2.9388 (4)	0.0070 (10)	0.0114 (6)	0.0012	0.0013	0.0004	0.0004
0.2	2.0673 (9)	2.7171 (5)	0.0147 (2)	0.0450 (1)	0.0052	0.0052	0.0018	0.0019
0.3	1.6964 (5)	2.4996 (8)	0.0170 (2)	0.1063 (2)	0.0108	0.0114	0.0038	0.0038
0.4	1.4146 (4)	2.2938 (4)	0.0100 (2)	0.1872 (3)	0.0183	0.0198	0.0066	0.0067
0.5	1.202 (1)	2.1046 (4)	−0.0007 (5)	0.2705 (3)	0.0274	0.0304	0.0102	0.0103
0.6	1.041 (1)	1.9329 (8)	−0.009 (1)	0.3412 (3)	0.0383	0.0424	0.0145	0.0147
0.7	0.930 (1)	1.7797 (6)	−0.013 (2)	0.3927 (5)	0.0523	0.0559	0.0194	0.0197
0.8	0.886 (1)	1.643 (3)	−0.004 (1)	0.427 (5)	0.0745	0.0697	0.0259	0.0263

[a] Data for the ^1S and ^3S states are from Schwartz (1961b), data for the $^1P^0$ and $^3P^0$ states are from Armstead (1968), data for the ^1D and ^3D states are from Register and Poe (1975), and data for the $^1F^0$ and $^3F^0$ states are from Callaway (1978a,b).

numbers $LS\pi$. For scattering from a ^2S target state, in partial wave l, $L = l$, $S = 0$ or 1, and π is $(-1)^l$.

Williams (1975) showed that phase shifts for partial waves with $l \geq 3$ must be included in the partial wave expansion in order to represent the e^-–H differential cross section accurately for scattering angles below 30°, for energies as great as 8.7 eV $(k = 0.8a_0^{-1})$. He used the partial wave Born formula for $l \geq 3$ with $\alpha_d = 4.50a_0^3$, the exact value for 1s hydrogen, together with theoretical values of η_0, η_1, and η_2 to compute the differential cross section. In general, the experimental data agreed within its error limits with results of theoretical models that had taken into account the full polarizability of ground-state hydrogen. These models include the variational calculations of Schwartz (1961b), Armstead (1968), and Gailitis (1965), the polarized orbital model (Temkin and Lamkin, 1961; Temkin, 1962), and close-coupling calculations that included a pseudostate to represent the full effect of α_d (Burke *et al.*, 1969b).

Castillejo *et al.* (1960) have analyzed the electric dipole polarization potential resulting from the close-coupling expansion. In the three-state (1s, 2s, 2p) approximation for atomic hydrogen, the effective value of α_d is 66% of its true value. The expansion in bound p states is slowly convergent. The sum over all bound states yields only 81% of the polarizability, leaving a contribution of 19% from the ionization continuum. In the static exchange approximation, the polarizability is zero, which makes this approximation inaccurate for the differential cross section at forward scattering angles. For angles less than 70°, the three-state close-coupling calculation also falls well outside the experimental error limits (Williams, 1975). This behavior appears to be corrected by including a pseudostate p-orbital $u_{1s \to p}$ in the close-coupling expansion. This function, which can be derived in closed form for hydrogen (Damburg and Karule, 1967), is

$$u_{1s \to p}(r) = \left(\frac{32}{129}\right)^{1/2}\left(1 + \frac{r}{2}\right)r^2 e^{-r}, \tag{5.3}$$

for a normalized p orbital whose radial factor is $r^{-1}u(r)$. It should be noted that this is a much more compact function than the 2p orbital, whose radial factor is proportional to $r \exp(-r/2)$.

The original s-wave calculations of Schwartz (1961b), using the Kohn method, were repeated and verified by Shimamura (1971a). Shimamura also used the single-channel method of Harris (1967), in which $\tan \eta$ is given by Eq. (2.50). The discrete set of energy values at which this method is valid was varied by means of a nonlinear parameter in the variational wave functions. For both ^1S and ^3S scattering states, with 56 and 50 term trial functions, respectively, the Harris values and Kohn values of phase shifts, computed at the same energies, were both in reasonable agreement with the

Kohn values extrapolated to convergence of the basis expansion. However, when given to four decimal places, the Harris values appear to be noticeably less accurate than the Kohn values. This is expected since the Harris formula is not stationary.

Similar calculations of s-wave phase shifts have been carried out, using a variational least-squares or minimum-norm method, by Abdel-Raouf and Belschner (1978). Variational bounds for these phase shifts were computed by Abdel-Raouf (1979).

The anomaly-free method (Nesbet, 1968, 1969a), described in Section 2.4, was used by Register and Poe (1975) to compute phase shifts for orbital s, p, and d waves. The calculations used Hylleraas (1928) basis functions. The computed d-wave phase shifts are included in Table 5.1. The s and p phase shifts were in good agreement with Schwartz and Armstead, respectively. These calculations by Register and Poe showed that the repetition of calculations with different values of a nonlinear parameter required in direct use of Kohn's method could be avoided by use of an anomaly-free method.

The single-channel method of Rudge (1973, 1975), which avoids anomalies as described in Section 2.4, was used to compute s-wave phase shifts for hydrogen, with results comparable to those of Schwartz. Similar calculations were carried out for the p-wave phase shifts (Das and Rudge, 1976).

Because they include the relative coordinate r_{12} explicitly, Hylleraas basis functions cannot easily be used for systems with more than two electrons. A viable approach for $(N + 1)$-electron systems is to use sums of product functions in the individual electronic coordinates, or Slater determinants as in Eqs. (1.6) and (1.7). Since the Hylleraas expansion is inappropriate to the simultaneous representation of two or more different target states, this second approach has been used for e^-–H scattering calculations above the $n = 2$ threshold (Callaway and Wooten, 1974).

Callaway (1978a) used programs developed for such calculations above the first excitation threshold to compute d- and f-wave phase shifts in the elastic scattering region. The two-electron variational basis functions were expressed as Slater determinants or their equivalents. The orbital basis set included functions that could represent the orbitals $1s$, $2s$, $2p$ and the pseudostate function $u_{1s \to p}$ exactly, together with other basis functions to represent short-range correlation effects and the inner part of open- and closed-channel orbital functions. The calculations by Callaway and Wooten (1974) were organized as an application of the matrix variational method, described here in Chapter 4. The f-wave phase shifts computed by Callaway (1978a) are included in Table 5.1. The computed d-wave phase shifts were in good agreement with those of Register and Poe (1975), who used a Hylleraas basis of up to 84 terms.

At least two independent variational calculations have been carried out for each of the e^-–H elastic scattering phase shifts with $l \leq 2$. The quasi-minimum principle of Hahn (1971), which can be based on Eq. (2.57), indicates that $\tan \eta$ should be greater for the more accurate calculation at given energy. This result is rigorous if error due to variational approximation of the equivalent close-coupling equations is negligible. The pattern of convergence found by Schwartz in calculations with the unmodified Kohn method is illustrated in Figs. 2.1 and 2.2. Except for anomalies, which apparently violate the conditions of the variational bound theorem, for a fixed value of the parameter κ, the computed values of $\tan \eta$ increase monotonically with the number of basis functions used. This essentially monotonic behavior was used by Schwartz (1961b) and Armstead (1968) in extrapolating their results to convergence.

A series of elastic electron–hydrogen scattering resonances occurs just below the $n = 2$ threshold. The first indication of these resonances was given by three-state close-coupling calculations (Burke and Schey, 1962). Gailitis and Damburg (1963b) showed that the degenerate $2s$ and $2p$ levels of hydrogen produce an effective static dipole potential, varying as r^{-2} for large r. If wave functions were represented in an orbital basis from which $1s$ is excluded, this would give an infinite series of states bound with respect to the $n = 2$ threshold. When allowed to interact with the $1s$ elastic scattering continuum, these states become resonances. The effective static dipole potential due to $2s$, $2p$ degeneracy results in an external orbital function of mixed parity. If the $l = 0$ and $l = 1$ projections of this function are denoted by $\bar{s}$ and $\bar{p}$, respectively, the ^1S and ^3S resonances considered by Chen (1967) are described by mixed configurations $2s\,\bar{s}$ and $2p\,\bar{p}$. The first member of the ^1S series must be strongly influenced by the excited valence state (VS) configurations $2s^2$ and $2p^2$. With reference to the qualitative discussion of resonances in Section 3.1, the higher members of these series are examples of nonvalence (NV) states since the outer electron does not experience a Coulomb potential.

The resonance or metastable states of H^- associated with the $n = 2$ threshold of hydrogen should include excited VS structures due to configurations $2s^2$, $2s2p$, and $2p^2$. In addition to the series of ^1S and ^3S resonances, mixed configurations $2s\,\bar{p}$ and $2p\,\bar{s}$ should give series of $^1P^0$ and $^3P^0$ resonances. The lowest $^3P^0$ resonance must be influenced by the VS configuration $2s2p$, while higher members of this series would be NV resonances. In the case of the $^1P^0$ series, the exchange integral between $2s$ and $2p$ will cause $(2s2p)^1P^0$ to lie above $(2s2p)^3P^0$, so the singlet state may lie above the $n = 2$ threshold or be strongly mixed with NV states. The $(2p^2)^3P$ state is metastable, since a state of even parity with $L = 1$ cannot interact with the electron scattering continuum of the $(1s)^2S$ hydrogen ground state if it lies below the $n = 2$ threshold at 10.204 eV. Calculations by Holøien (1960) and by Drake (1970) indicate that the energy of this state is 10.200 eV. The pure state $(2p^2)^1D$ is unlikely to lie below $(2p^2)^3P$ but may contribute to a ^1D series of NV resonances arising from mixed configurations $2s\,\bar{d}$ and $2p\,\bar{p}$. Close-coupling calculations that include states beyond $n = 2$ (Burke *et al.*, 1967) or that include electron correlation terms (Taylor and Burke, 1967) locate a $^1P^0$ resonance in the inelastic scattering region just above the $n = 2$ threshold. This resonance may be associated with the $2s2p$ configuration of H^-, but a detailed analysis of the resonance state has not been carried out. Close-coupling calculations (Burke, 1968) locate a $^1P^0$ resonance at 10.177 eV, below the $n = 2$ threshold, and a series of NV resonances is possible.

The theoretical studies of e^-–H resonances have been reviewed by Burke (1968), by Chen (1970), and by Callaway (1978b). Schulz (1973a) compares theoretical and experimental data as of 1972. For the lowest-lying resonances, variational scattering calculations of resonance energies and widths have been carried out at the level of accuracy of the corresponding phase-shift calculations. Table 5.2 compares results of those variational calculations expected to be the most accurate, as indicated by the variational bound property of the corresponding phase shifts, with experimental data (McGowan, 1967; Sanche and Burrow, 1972). The three resonances considered are those that are best characterized experimentally (Schulz, 1973a). In each case, theory and experiment agree within the experimental error estimates. The theoretical calculations are those of Shimamura (1971b), using Kohn's method; of Das and Rudge (1976), using the anomaly-free method of Rudge (1973); and of Register and Poe (1975), using the anomaly-free method of Nesbet (1968).

The character of the resonances listed in Table 5.2 as short-lived states of H^- has been confirmed experimentally by Risley *et al.* (1974), who observed the electrons emitted by autodetachment from H^- excited states

Table 5.2. Variational Calculations of Resonances in e^-–H Elastic Scattering

State	Energy (eV)		Width (eV)	
^1S	9.557^a	9.558 ± 0.01^d	0.04720^a	0.0430 ± 0.0060^e
$^3P^0$	9.738^b	9.738 ± 0.01^d	0.00585^b	0.0056 ± 0.0005^d
^1D	10.122^c	10.128 ± 0.01^d	0.00900^c	0.0073 ± 0.0020^d

[a] Shimamura (1971b).
[b] Das and Rudge (1976).
[c] Register and Poe (1975).
[d] Sanche and Burrow (1972).
[e] McGowan (1967).

produced by ion–atom collisions. The observed energies of emitted electrons corresponded to H$^-$ states of energy 9.59(3) eV, 9.76(3) eV, and 10.18(3) eV, respectively, in good agreement with the data of Table 5.2.

The lowest ^1S, $^3P^0$, and ^1D resonances are probably of VS character and should be well represented by variational calculations with Hylleraas basis functions. Variational calculations using exponential basis orbitals in the three-state algebraic close-coupling approximation were carried out by Seiler *et al.*, (1971). These calculations obtained several NV resonances in each of the series ^1S, ^3S, $^1P^0$, and $^3P^0$. The ^3S and higher ^1S resonance energies and widths obtained in this approximation agreed closely with the model calculations of Chen (1967). The calculations of Chen indicate that a model Hamiltonian can easily be constructed for the long-range part of the effective dipole potential due to the $2s$, $2p$ degeneracy. Accurate calculations of higher members of the NV resonance series could probably be carried out by using such a model Hamiltonian in one of the hybrid methods discussed in Section 2.6.

5.2. Hydrogen: Inelastic Scattering

The most extensive calculations of e^-–H scattering above the $n = 2$ threshold have been carried out by Callaway and collaborators using variational methods. This work has been reviewed in detail by Callaway (1978b). The range of incident energies considered is 10.2 to 54.4 eV, well above the ionization threshold at 13.6 eV. Ground-state elastic scattering and excitation to the $n = 2$ states have been considered.

Above the $n = 2$ threshold, an effective static dipole potential exists in the $2s$ and $2p$ open channels. This requires modification of the asymptotic form of continuum basis functions used in variational calculations. Otherwise the variational expansion of the scattering wave function becomes very slowly convergent. Callaway and collaborators introduced energy-dependent quadratically integrable basis functions in the form of oscillatory functions divided by powers of r at large r (Seiler *et al.*, 1971).

The variational calculations were organized in the algebraic close-coupling formalism. Target hydrogen atom bound and continuum states were represented by a pseudostate expansion that included $1s$, $2s$, and $2p$ orbitals and individual components of the polarization function $u_{1s \to p}$ given by Eq. (5.3), together with functions chosen for their effect on short-range correlation (Matese and Oberoi, 1971; Callaway and Wooten, 1974, 1975). Only the $1s$, $2s$, and $2p$ channels were treated as open channels, with appropriate continuum basis functions. Quadratically integrable basis functions for both open and closed channels were represented by exponential functions, including functions with very small exponents required for an adequate representation of the outer part of closed-channel orbitals. Results were obtained with several variants of the Kohn formalism, as discussed in Section 2.4, and then compared to give an estimate of the error due to choice of formalism. Integrals involving oscillatory basis functions were evaluated analytically (Oberoi *et al.*, 1972).

Convergence of the variational expansion could be checked in the elastic scattering region by the lower-bound property of the phase shifts. In the region between the $n = 2$ and $n = 3$ thresholds, a similar lower-bound principle is valid for the sum of eigenphases since the $2s$ and $2p$ target states are represented exactly. Typical results of such checks and comparison with calculations by other methods are given by Callaway (1978b). These comparisons indicate that the variational calculations above the $n = 2$ threshold are the most accurate available and are comparable to calculations found to be in quantitative agreement with experiment in the elastic scattering region.

Variational calculations were carried out only for total $L \leq 3$. In order to compute accurate differential cross sections and total excitation cross sections at higher energies, higher angular momentum contributions to the K matrix were obtained by use of the distorted-wave polarized orbital method (DWPO) (McDowell *et al.*, 1974) and of the unitarized Born approximation (Callaway *et al.*, 1975). The hydrogen ground-state dipole polarizability α_d and quadrupole polarizability α_q, together with the coefficient β of the leading term in the nonadiabatic correction to the dipole polarization potential, contribute in the partial-wave Born approximation to an elastic scattering phase shift η_l in the form (Callaway, 1978b), for $l \geq 1$,

$$\tan \eta_l = \frac{\pi \alpha_d k^2}{(2l + 3)(2l + 1)(2l - 1)} \left[1 - \frac{9b^2 k^2}{(2l + 5)(2l + 3)} + O(k^4) \right], \quad (5.4)$$

where

$$b^2 = (\beta - \alpha_q)/3\alpha_d = 23/18 \quad (5.5)$$

for the hydrogen ground state. This was used for large orbital l values, while an effective potential

$$V_{\text{eff}}(r) = -\alpha_d r^2/(r^2 + b^2)^3 \quad (5.6)$$

that reproduces Eq. (5.4) was used for smaller l values. The resulting value of $\tan \eta_l$ then replaced the element $K_{1s,1s}$ of the K matrix in the unitarized Born approximation.

These calculations obtained elastic and $2s$, $2p$ excitation cross sections in the energy range between the $n = 2$ and $n = 3$ thresholds (Callaway and Wooten, 1974; Callaway *et al.*, 1975, 1976; Morgan *et al.*, 1977).

The $^1P^0$ resonance just above the $n = 2$ threshold was located in these calculations at $E = 10.225$ eV, with width $\Gamma = 0.014$ eV, in good agreement with resonance parameters $E = 10.222$ eV, $\Gamma = 0.015$ deduced by Macek and Burke (1967) from the close-coupling calculations of Taylor and Burke (1967). Resonance structure in the $2s$ and $2p$ excitation cross sections within 0.04 eV of the $n = 2$ threshold has been observed by McGowan *et al.* (1969) and by Williams and Willis (1974).

Several series of resonances occur just below the $n = 3$ excitation threshold, analogous to the elastic scattering resonances below the $n = 2$ threshold. The energies and widths of many of these resonances were computed by Morgan *et al.* (1977). The computed resonance energies are indicated in Figs. 5.1 and 5.2, which show the total excitation cross sections for $1s \rightarrow 2s$ and $1s \rightarrow 2p$, respectively (Callaway, 1978b). The theoretical calculations are compared with experimental data points of Williams (1976). Typical experimental error bars are shown in both figures. The figures show quantitative agreement between experiment and theory, except for details of the complex resonance structure.

Above the $n = 3$ threshold, the pseudostate expansion used by Callaway and collaborators cannot represent the level structure of hydrogen in detail. In the intermediate energy region, above the ionization threshold at 13.6 eV, no true resonance structure is expected, but the assumed wave function cannot represent the two-electron ionization continuum. Ioniza-

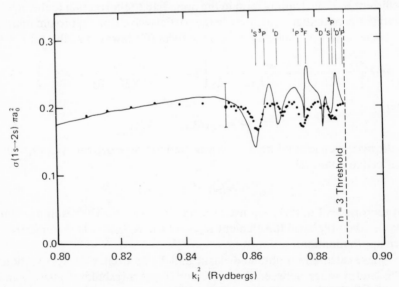

Figure 5.1. e^-–H, total $1s \rightarrow 2s$ excitation cross section (Callaway, 1978b, Fig. 4).

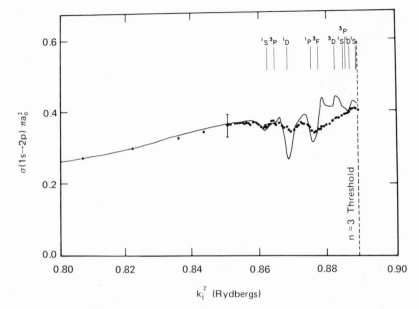

Figure 5.2. e^-–H, total $1s \rightarrow 2p$ excitation cross section (Callaway, 1978b, Fig. 5).

tion is approximated crudely as excitation of discrete pseudostates with energies greater than 13.6 eV. It is assumed that elastic scattering and excitation of low-lying states can be described adequately so long as the rate of electron-impact ionization is small.

The inherent difficulty with this approach is that the pseudostates, acting as if they were true bound states, are accompanied by pseudoresonances and threshold structures. Unlike the anomalies in the Kohn variational method, discussed in Section 2.3, these spurious features are inherent in the physical model that replaces the ionization continuum by discrete states. In calculations for energies up to 54.4 eV, Callaway *et al.* (1976) adjusted parameters in the variational basis orbital set so that pseudostate energies and the accompanying pseudoresonance structures would fall in between the scattering energy values used. It should be emphasized that no obviously better solution has yet been found to the formal and practical difficulties inherent in the theory of intermediate energy scattering.

Results of these calculations are shown in Figs. 5.3 and 5.4 (Callaway, 1978b). Figure 5.3 compares the computed differential elastic cross section in the energy range 12.25 eV ($k^2 = 0.90\ a_0^{-2}$) to 30.61 eV ($k^2 = 2.25\ a_0^{-2}$) with experimental data of Williams (Callaway and Williams, 1975). The general agreement is good.

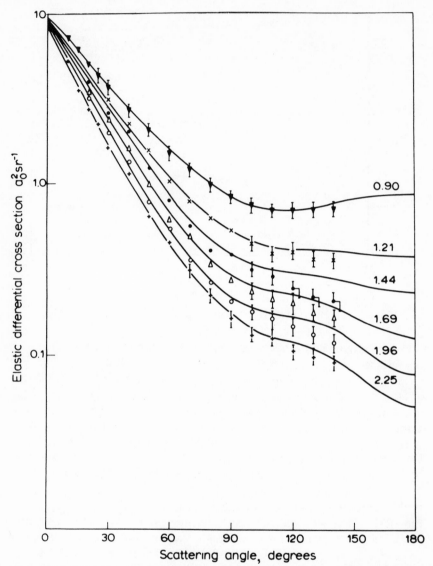

Figure 5.3. e^-–H, differential elastic cross section above the $n = 3$ threshold. Theoretical values (solid lines) are compared with experimental data points (Callaway, 1978b, Fig. 9).

Figure 5.4 compares the computed differential cross section for excitation of the $n = 2$ states with data of Williams (1976) at energy 19.59 eV ($k^2 = 1.44\ a_0^{-2}$). The agreement between theory and experiment appears to be excellent.

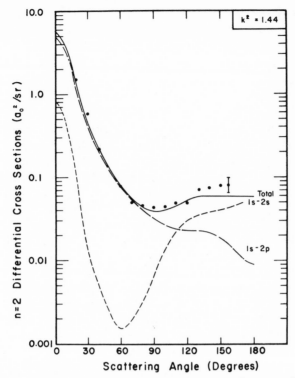

Figure 5.4. $e^- - H$, differential $2s + 2p$ excitation cross section at $k^2 = 1.44\,a_0^{-2}$, or 19.59 eV. Theoretical values (solid and dashed lines) are compared with experimental data points (Callaway, 1978b, Fig. 10).

5.3. Alkali Metal Atoms

Because of the strong interaction between their n_0s and n_0p valence states, the alkali metal atoms have very large ground-state electric dipole polarizabilities α_d. The values recommended in a review by Miller and Bederson (1977) are

$$\text{Li:} \quad \alpha_d = 164\,a_0^3,$$

$$\text{Na:} \quad \alpha_d = 159\,a_0^3, \tag{5.7}$$

$$\text{K:} \quad \alpha_d = 293\,a_0^3,$$

to be compared with $4.5\,a_0^3$, the ground-state electric dipole polarizability of hydrogen. These very large polarizabilities make it desirable to solve the full continuum two-electron problem for the valence and external electrons as accurately as possible. Mittleman (1966) proposed that this problem could be solved as a continuum Bethe–Goldstone equation. Matrix variational calculations by Sinfailam and Nesbet (1973), to be discussed here, used this formalism, as described in Section 1.5.

A short review of theory and experimental data on low-energy electron scattering by alkali metal atoms has been given by Bederson (1970a,b). Bederson and Kieffer (1971) give a detailed critical review of relevant experimental techniques. The conclusion drawn from theoretical calculations and from modern experiments is that in early measurements by Brode (1929) the absolute e^-–K cross section was roughly two times too large at all energies and showed spuriously exaggerated structure at the first excitation threshold. Since no comparable absolute experimental data were available until recently, there was a long-standing discrepancy between theory and experiment.

In the elastic scattering region, electron scattering by an alkali metal atom is dominated by the large electric dipole polarizability. The oscillator strength of the $n_0s \rightarrow n_0p$ optical transition is nearly unity, and the ground-state polarizability is almost entirely accounted for by the transition dipole moment between these two states. In these circumstances, the two-state (n_0s and n_0p) close-coupling approximation should give excellent results for elastic cross sections. In contrast to the case of hydrogen, where a pseudo-state $u_{1s \rightarrow p}$ must be included in addition to the physical $2p$ function, the physical n_0p orbital for an alkali metal atom suffices as a polarization pseudostate.

Two-state close-coupling calculations were carried out by Karule (1965, 1972) for elastic electron scattering by Li, Na, K, and Cs. Spin polarization and differential elastic cross sections were computed from the close-coupling phase shifts. A two-state $(2s, 2p)$ close-coupling calculation of e^-–Li scattering, extending into the inelastic region, was carried out by Burke and Taylor (1969). The elastic partial-wave phase shifts were in agreement with those of Karule, except for a sharp peak just above threshold in the ^3S phase shift of Burke and Taylor, similar to a structure that might arise from a virtual state. Norcross (1971) repeated the ^3S calculation with corrections for orthogonalization effects and showed that this peak was a computational artifact.

Two-state close-coupling calculations were carried out by Norcross (1971) for both Li and Na using an effective potential to represent the atomic ion core. Norcross introduced a refined computational technique to improve the accuracy of calculations near threshold. This work was extended to the inelastic region for Na (up to 5 eV), including study of the effect of additional states $(4s, 3d)$ in the close-coupling expansion (Moores and Norcross, 1972). Variational Bethe–Goldstone calculations by Sinfailam and Nesbet (1973), using the OAF (optimized anomaly-free) method (Nesbet and

Oberoi, 1972), confirmed these close-coupling results, which are in substantial internal agreement when corrected for the computational difficulties analyzed by Norcross (1971). Vo Ky Lan (1971) has shown that the polarized orbital model, including the exchange polarization potential, can reproduce the two-state close-coupling results for e^-–Li elastic scattering.

Figures 5.5, 5.6, and 5.7 show computed values of the three lowest-order partial-wave phase shifts, both singlet and triplet, for Li (Sinfailam and

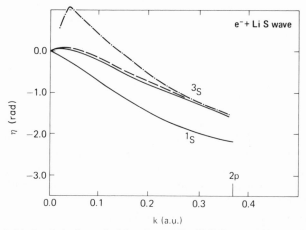

Figure 5.5. e^-–Li, $L = 0$ singlet and triplet phase shifts (Sinfailam and Nesbet, 1973, Fig. 1). Curves: ——, Sinfailam and Nesbet (1973); –·–, Burke and Taylor (1969); – –, Norcross (1971).

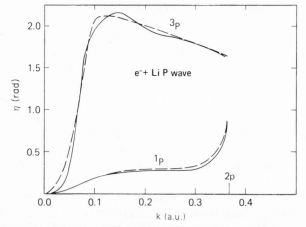

Figure 5.6. e^-–Li, $L = 1$ singlet and triplet phase shifts (Sinfailam and Nesbet, 1973, Fig. 2). Curves: ——, Sinfailam and Nesbet (1973); – –, Burke and Taylor (1969).

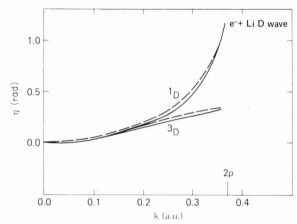

Figure 5.7. e^-–Li, $L = 2$ singlet and triplet phase shifts (Sinfailam and Nesbet, 1973, Fig. 3). Curves: ——, Sinfailam and Nesbet (1973); – –, Burke and Taylor (1969).

Nesbet, 1973). The orbital basis augmented a double-zeta representation of the ground state and excited $2p$ Hartree–Fock orbitals (Nesbet, 1970) with exponential basis functions whose exponents form a geometric sequence. The variational calculations are compared in the figures with two-state close-coupling calculations (Burke and Taylor, 1969). The corrected ^3S phase shift computed by Norcross (1971) is included in Fig. 5.5. With this correction, the close-coupling and variational phase shifts are in close agreement.

Figures 5.8, 5.9, and 5.10 show a similar comparison for Na. Variational Bethe–Goldstone calculations (Sinfailam and Nesbet, 1973) give e^-–Na elastic phase shifts in substantial agreement with close-coupling results (Norcross, 1971; Moores and Norcross, 1972).

For a given partial wave, the phase shifts of Li and Na are very similar functions of k. This similarity holds for all of the alkali metal atoms. The ^1S phase shift descends linearly from its threshold value (which could be taken to be π radians) with negative initial slope, due to the ^1S ground state of the negative ion lying below the threshold, in accord with the analytical theory discussed in Section 3.3. The electronic configuration of this negative ion state is $(n_0 s)^2$, with a large admixture of $(n_0 p)^2$. The ^3S phase shift rises from threshold with positive slope, then passes through a low maximum and descends smoothly.

The $^3P^0$ phase shift rises from threshold, following a resonance curve that is modified in shape because the threshold law, Eq. (3.99), requires a p-wave phase shift to vary as k^3 for small k. In the variational calculations of Sinfailam and Nesbet (1973), the resonance search procedure (Nesbet and

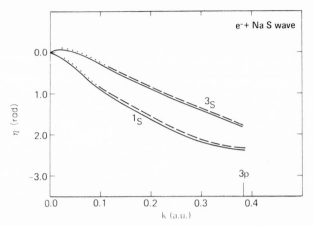

Figure 5.8. e^-–Na, $L = 0$ singlet and triplet phase shifts (Sinfailam and Nesbet, 1973, Fig. 4). Curves: ——, Sinfailam and Nesbet (1973); – –, Moores and Norcross (1972); · · ·, Norcross (1971).

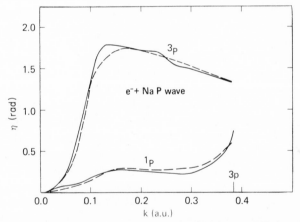

Figure 5.9. e^-–Na, $L = 1$ singlet and triplet phase shifts (Sinfailam and Nesbet, 1973, Fig. 5). Curves: ——, Sinfailam and Nesbet (1973); – –, Moores and Norcross (1972).

Lyons, 1971), described in Section 3.2, gave the following values for the resonance energy E_{res} and width Γ of the $^3P^0$ resonances:

$$\text{Li}^-: \quad E_{res} = 6.0 \times 10^{-2}\,\text{eV}, \quad \Gamma = 5.7 \times 10^{-2}\,\text{eV},$$

$$\text{Na}^-: \quad E_{res} = 8.3 \times 10^{-2}\,\text{eV}, \quad \Gamma = 8.5 \times 10^{-2}\,\text{eV}, \quad (5.8)$$

$$\text{K}^-: \quad E_{res} = 2.4 \times 10^{-3}\,\text{eV}, \quad \Gamma = 5.8 \times 10^{-4}\,\text{eV}.$$

A striking implication of the threshold law, as indicated in the discussion

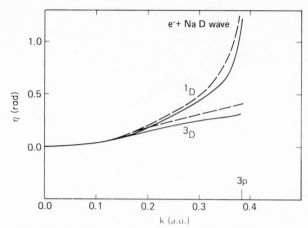

Figure 5.10. e^-–Na, $L = 2$ singlet and triplet phase shifts (Sinfailam and Nesbet, 1973, Fig. 6). Curves: ——, Sinfailam and Nesbet (1973); – –, Moores and Norcross (1972).

following Eq. (3.86), is that if a model potential is varied so that k_{res}^2 approaches zero, the resonance width vanishes as k_{res}^{2l+1}. The resonance is effectively squeezed against the threshold. The present $^3P^0$ resonances provide an example of this behavior. In particular, the K^- resonance, very close to the threshold, becomes extremely narrow. These resonances correspond to $(n_0 s, n_0 p)^3 P^0$ excited valence states of the negative ions, analogous to these states in the isoelectronic alkaline earth atoms. These are shape resonances in the usual terminology but are very narrow because of their proximity to the scattering threshold. Because of difficulties in defining low-energy electron beams, there are no reliable experimental data close enough to the elastic threshold to identify these resonances.

The $^1P^0$ and ^1D phase shifts both rise sharply below the $n_0 p$ excitation threshold. As shown for Li in Fig. 5.11 (Bardsley and Nesbet, 1973), there is a ^1D resonance peak in the elastic cross section at or near the $2p$ threshold, and a Wigner cusp is superimposed on a $^1P^0$ resonance peak, which may be associated with the $(n_0 s, n_0 p)^1 P^0$ excited valence state of Li$^-$. The state $(n_0 p, n_0 p)^1 D$ contributes to the ^1D resonance. Figure 5.11 shows partial-wave contributions to the elastic scattering cross section, computed by the matrix variational method, near the $2\,^2P^0$ threshold, whose calculated value was 1.8411 eV. The cross sections in total S and D states pass smoothly through the threshold. There is a prominent cusp in the $^1P^0$ partial cross section, and the $^3P^0$ cross section shows a small but real step at the threshold.

As shown in Section 3.3, cusp behavior occurs at the excitation threshold of a state with given L_γ only for partial waves with orbital $l_p = 0$ in the new channel. This affects only those components of the total wave

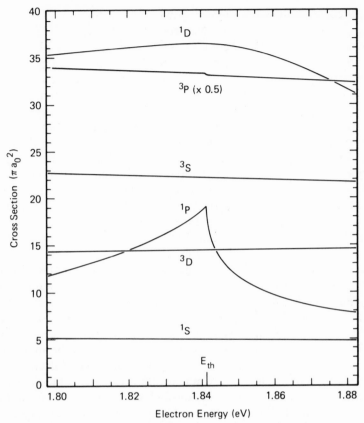

Figure 5.11. e^-–Li, partial wave contributions to the elastic cross section near the $2p$ excitation threshold (Bardsley and Nesbet, 1973, Fig. 1).

function with $L = L_\gamma$. In the present case, at the $(n_0 p)^2 \, {}^2 P^0$ threshold, cusp behavior occurs only in the ${}^1 P^0$ and ${}^3 P^0$ scattering states and K matrices. The analytic theory implies that the ${}^1 P^0$ cross section shown in Fig. 5.11 should have infinite slope on both sides of the threshold. The finite slope shown below threshold is probably due to an inadequate trial wave function (Bardsley and Nesbet, 1973). The basis set did not include any terms (energy-dependent exponents in exponential functions) with exactly the correct asymptotic behavior for closed channels. Above threshold, continuum functions with the correct asymptotic form are included, so the correct analytic form is obtained. A better procedure would be to use Eq. (3.109), or the specific forms of Eqs. (3.121) and (3.122) in the present case, to deduce the cross section just below threshold from that computed just above.

The corresponding threshold cusp in the $^1P^0$ partial elastic cross section has been computed for Na by Norcross and Moores (1973), who also computed the photodetachment cross section of Na$^-$. Since photodetachment proceeds by optical absorption from the ^1S negative ion ground state, the $^1P^0$ continuum state is singled out by optical selection rules. The photodetachment cross section (Norcross and Moores, 1973) shows the $^1P^0$ cusp as a prominent feature. Since the np_0 threshold energy is known spectroscopically to high precision, observation of the photodetachment cusp provides a precise measurement of the electron affinity of the neutral atom. Patterson *et al.* (1974) have used this technique to obtain accurate experimental values of electron affinities of the alkali metal atoms.

The elastic phase shifts computed variationally and in the two- or four-state close-coupling approximation are in close agreement, as shown in Figs. 5.5–5.10. Total cross sections for scattering by Li, Na, and K are shown in Figs. 5.12–5.14, respectively, and are compared with experimental data. Figure 5.12 shows the variational Bethe–Goldstone calculation (Sinfailam and Nesbet, 1973) as a full line, together with the close-coupling calculations by Norcross (1971) and Karule (1965, 1972). The experimental points are from Perel *et al.* (1962). Figure 5.13, from the experimental paper by Kasdan *et al.* (1973), shows their experimental data for e^-–Na scattering compared with two-state close-coupling calculations of Karule and Peterkop (1965) and of Moores and Norcross (1972). Within the range of the experimental points, for elastic scattering, the variational calculations and the calculations of Moores and Norcross follow essentially the same curve. The latter calculations continue into the inelastic scattering region and show cusp structure at the inelastic threshold, as discussed above. Figure 5.14 shows the e^-–K elastic cross section as computed variationally (full line: Sinfailam and Nesbet, 1973), compared with the close-coupling calculation of Karule (1965, 1972) and with experimental data of Collins *et al.* (1971).

Figure 5.15 shows data, as in Fig. 5.14, for the elastic spin-exchange cross section. This can be derived from Eqs. (1.30) and (1.31) as

$$\sigma^{SE} = \frac{\pi}{k^2} \sum_L (2L + 1) \sin^2 (\eta_{L0} - \eta_{L1}), \qquad (5.9)$$

in terms of the singlet and triplet phase shifts, η_{L0} and η_{L1}, respectively.

In the experiment of Collins *et al.* (1971), a crossed electron and atomic beam recoil technique was used, with velocity selection and spin selection and analysis of the atomic beam, to measure e^-–K differential cross sections with and without spin exchange. The absolute total cross section is measured in the same experiment. The resulting experimental total cross section and spin-exchange cross section are shown in Figs. 5.14 and 5.15, respectively.

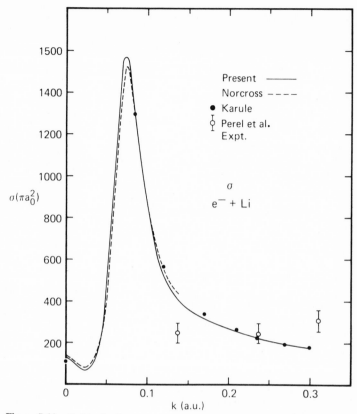

Figure 5.12. e^-–Li, elastic cross section (Sinfailam and Nesbet, 1973, Fig. 10).

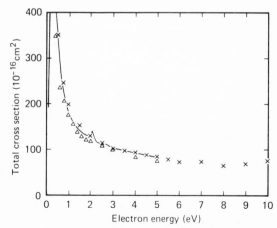

Figure 5.13. e^-–Na, total cross section (Kasdan *et al.*, 1973, Fig. 4). Data: ×, experiment; △, Karule and Peterkop (1965); ——, Moores and Norcross (1972).

The comparison with theoretical calculations is good within the experimental error bars for σ^{SE}, but the experimental data do not extend to low enough energies to probe the resonance peak predicted by theory.

In the case of sodium, the experimental data of Kasdan *et al.* (1973), shown in Fig. 5.13, show a strong rise at low energies, following the theoretical curve, but do not extend as low as the resonance peak.

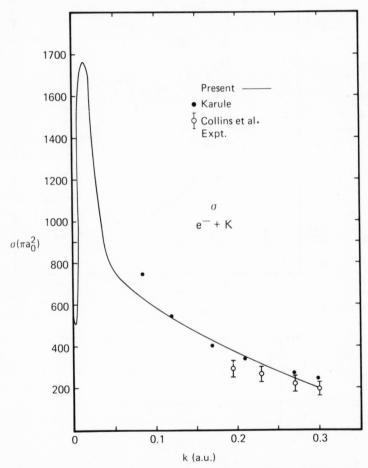

Figure 5.14. e^-–K, elastic cross section (Sinfailam and Nesbet, 1973, Fig. 14).

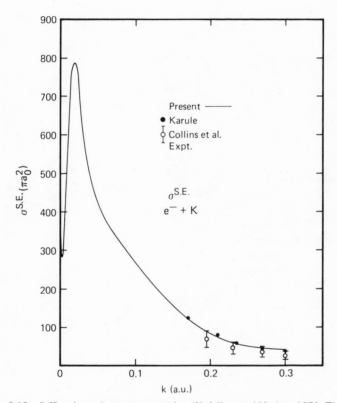

Figure 5.15. e^-–K, spin-exchange cross section (Sinfailam and Nesbet, 1973, Fig. 15).

6

Applications to Other Atoms

Introduction

For atoms beyond hydrogen, there are no exact wave functions. Inter-electronic relative coordinates provide very accurate variational wave functions for helium. Such coordinates cannot easily be used even for the three-electron continuum problem of electron–helium scattering, and they present great difficulties for heavier atoms. Some compromise must be made in the accuracy of representation of a complex target atom in order to make calculations of electron scattering feasible. Even for light atoms, most existing calculations have been limited to the Hartree–Fock approximation for target atom states, while the polarization response is represented at the level of first-order perturbation theory.

In both theory and experiment, the most thoroughly studied complex target atom is helium. Variational calculations of elastic and inelastic scattering have been carried out at several levels of accuracy and have been compared with existing experimental data. In the elastic scattering range, variational calculations have established a standard differential cross section that can be used to calibrate experimental apparatus.

Electron scattering by the light open-shell atoms C, N, and O is important in astrophysics and atmospheric physics but is very difficult to study in the laboratory. Variational calculations predict structural features and quantitative values of cross sections and excitation rates. New experimental work is needed to confirm or to refine these predictions.

Theory and calculations of low-energy electron scattering by complex atoms have been reviewed by Nesbet (1977a). Low-energy electron scattering and attachment by C, N, and O atoms are reviewed by LeDourneuf et al. (1977) and by Nesbet (1977c).

6.1. Helium: Elastic Scattering

The matrix variational method was first applied to e^-–He scattering by Michels *et al.* (1969), with detailed results for s, p, and d elastic phase shifts reported by Harris and Michels (1971). This work was extended (Sinfailam and Nesbet, 1972) by carrying the orbital basis expansion to practical completeness within the framework of the Bethe–Goldstone approximation of level [1s] for He, as defined in Section 1.6. In principle, this approximation takes into account all polarization effects due to virtual excitation of a single target 1s orbital. Different ways of choosing basis orbitals with radial factors $r^n \exp(-\zeta r)$ were considered. The most satisfactory used a decreasing geometric sequence of exponents ζ augmented by an increasing arithmetic sequence. Convergence with respect to these sequences appeared to be approximately 0.1% in the phase shifts. The orbital l values included in the basis extended only to terms sufficient to give an accurate representation of electric dipole polarization effects for s, p, and d partial waves. The OAF (optimized anomaly-free) modification of Kohn's variational method was used (Nesbet and Oberoi, 1972).

The computed phase shifts are shown in Figs. 6.1, 6.2, and 6.3 as solid curves (present BG). For comparison, similar calculations were carried out at level [0] of the Bethe–Goldstone hierarchy, equivalent to the static exchange approximation. These results are shown as dashed curves (present SE). The static exchange approximation neglects the polarization potential, which is fully included at the target atom Hartree–Fock level in the [1s] or BG approximation.

Comparison with results obtained by different methods, also shown in Figs. 6.1, 6.2, and 6.3, verifies this theoretical interpretation. In each case, the BG results are nearly identical with calculations by variants of the polarized orbital method, which is designed to include electric dipole polarizability correctly. The SE variational results agree closely with prior static exchange calculations in the close-coupling formalism.

Figures 6.1–6.3 include phase shifts computed by the extended polarized orbital method (EPOM) (Callaway *et al.*, 1968) and static exchange calculations carried out with the R-matrix method (RM) (Burke and Robb, 1972). Figure 6.1 also includes the s-wave phase shift computed in the five-state close-coupling approximation by Burke *et al.* (1969a). The five states are 1^1S, 2^3S, 2^1S, 2^3P^0, and 2^1P^0 of helium. Because a polarization pseudostate is not included, this approximation cannot adequately represent the polarizability of the $1s^2$ target state. Full polarized orbital calculations (POM) (Duxler *et al.*, 1971) of the p- and d-wave phase shifts gave results shown in Figs. 6.2 and 6.3.

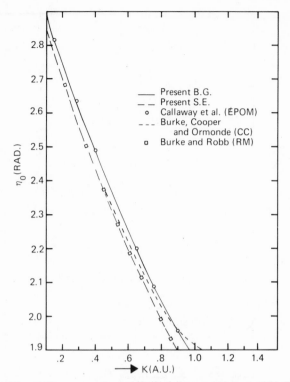

Figure 6.1. e^-–He, s-wave phase shift (Sinfailam and Nesbet, 1972, Fig. 5).

In the calculations of Burke *et al.* (1969a), an exchange term was approximated in a way that greatly exaggerated the 2^3S and 2^1S excitation cross sections. The effect of this approximation on elastic scattering is not known. The low-energy peak in the elastic cross section shown in Fig. 6.4 is attributed (Burke *et al.*, 1969a) to inclusion of only 35% of target atom polarizability in the five-state close-coupling approximation. Algebraic five-state close-coupling calculations were carried out by Wichmann and Heiss (1974). Because of the polarizability approximation, the computed phase shifts fall in between the static exchange and Bethe–Goldstone values.

Calculations of phase shifts for e^-–He elastic scattering were carried out by the GRPA coupled Green's function method (Yarlagadda *et al.*, 1973). This method provides an internally consistent approximate theory with an implied physical model similar to that of the polarized orbital or Bethe–Goldstone methods. The computed s and p phase shifts are in close agreement with polarized orbital and Bethe–Goldstone values. The computed d-wave phase shift was underestimated due to failure to include f

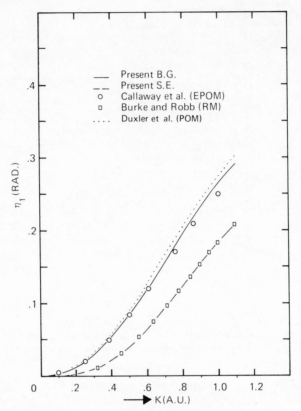

Figure 6.2. e^-–He, p-wave phase shift (Sinfailam and Nesbet, 1972, Fig. 6).

orbitals in the basis set used to represent the GRPA optical potential (Nesbet, 1975a).

The total elastic cross section computed in the matrix variational Bethe–Goldstone approximation (present work) (Sinfailam and Nesbet, 1972) is shown in Fig. 6.4, corrected from the originally published figure, which was misplotted. The EPOM total cross section, also shown in Fig. 6.4 (Callaway *et al.*, 1968), is nearly identical. The low-energy peak shown in the results of Burke *et al.* (1969a) is either a computational artifact or a consequence of an underestimate of target atom polarizability, as discussed above. The theoretical curves (BG and EPOM) are consistently above the upper error limits (3% estimated probable error) of the experiment of Golden and Bandel (1965), which was an absolute measurement of the scattering cross section.

The momentum-transfer cross section σ_M, which weights the differential cross section by a factor $(1 - \cos \theta)$, was also computed from the

Bethe–Goldstone phase shifts (Sinfailam and Nesbet, 1972) and was compared with experimental data of Crompton *et al.* (1967, 1970). This experiment, using a swarm technique, was estimated to give the absolute momentum-transfer cross section within error limits ±2% for energies below 2 eV and ±5% for energies up to 6 eV. The computed momentum-transfer cross section was just below the 2% error bars and within the 5% error bars. The small residual error was attributed to use of an uncorrelated target wave function (Sinfailam and Nesbet, 1972).

These results confirmed an inconsistency between the error estimates assigned to the absolute total cross-section measurements of Golden and Bandel (1965) and to the absolute momentum-transfer cross-section measurements of Crompton *et al.* (1967, 1970). Both cross sections are sums of positive terms coming from a small number of partial waves, and the theoretical cross section could not easily lie above the error limits in one case and below them in the other. In reviewing the experimental data, Bederson and Kieffer (1971) concluded that the total elastic cross section could be established to no better than 10–15% accuracy below 19 eV because of this

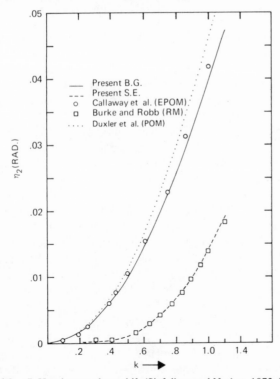

Figure 6.3. e^-–He, d-wave phase shift (Sinfailam and Nesbet, 1972, Fig. 7).

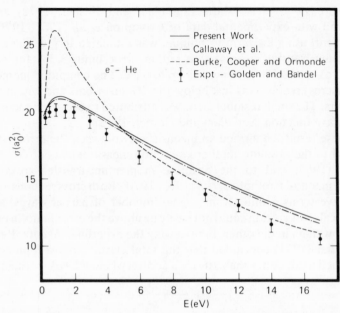

Figure 6.4. e^-–He, total cross section (Sinfailam and Nesbet, 1972, Fig. 10, corrected).

inconsistency. They used phase-shift analysis to compute an implied total cross section from the momentum-transfer cross section. This implied total cross section was found to be 9% greater than that measured by Golden and Bandel, well outside 3% error bars.

The experimental situation was clarified by Andrick and Bitsch (1975) who measured angular distributions to better than 5% accuracy and fitted their data to partial-wave phase shifts, which were used to compute total and momentum-transfer cross sections (Andrick, 1973). Following this work, it appeared that the general agreement between experiment and theory (BG or polarized orbital) was in the range of 5% or better and that swarm and beam data agreed within estimated error limits. New absolute total cross-section measurements by Kauppila and collaborators (Kauppila *et al.*, 1977; Stein *et al.*, 1978) and by Kennerly and Bonham (1978) confirmed this result and were consistent with the narrow error limits assigned to the earlier swarm data (Crompton *et al.*, 1967, 1970), which had been extended up to 12 eV (Milloy and Crompton, 1977).

A narrow resonance of ^2S symmetry occurs in e^-–He elastic scattering near 19.36 eV (Schulz, 1973a). The variational Bethe–Goldstone calculations of Sinfailam and Nesbet (1972) located this resonance at 19.42 eV. The background phase shifts were $\eta_0 = 104.9°$, $\eta_1 = 18.1°$, $\eta_2 = 3.2°$ (Nesbet, 1975a) in substantial agreement with values $\eta_0 = 105°$, $\eta_1 = 18°$,

$\eta_2 = 3.2°$ deduced from resonance scattering (Andrick, 1973). The variational calculations indicated that there is only this one resonance below the $n = 2$ excitation threshold. A search for other resonances in partial-wave channels with $l \leq 3$ showed none below the 2^3S threshold, and there were no indications of other energy-dependent structures. Similar conclusions had been reached by Temkin et al. (1972), who carried out resonance calculations using the projection-operator formalism of Feshbach. Sanche and Schulz (1972), reviewing tentative experimental observations of other structures in this energy range, concluded that structure found near 19.5 eV was an experimental artifact.

From observations of threshold structures in e^-–He elastic scattering at the 2^3S and 2^1S excitation thresholds, 19.818 and 20.614 eV, respectively, Cvejanovic et al. (1974) calibrated the ^2S resonance energy at 19.367 ± 0.008 eV. The width measured by Gibson and Dolder (1969) and by Golden and Zecca (1971) is 0.008 eV. There is an unresolved discrepancy between this value and the results of variational calculations. Temkin et al. (1972) found $\Gamma = 0.0144$ eV at $E_{res} = 19.363$ eV, while Sinfailam and Nesbet (1972) found $\Gamma = 0.015$ eV at $E_{res} = 19.42$ eV. These variational calculations neglect electronic correlation in the helium ground state. Since correlation must decrease the coefficient of the dominant $1s^2$ configuration, it will tend to reduce the integrals in Eq. (3.43) that determine the width of the ^2S resonance. The dominant configurations $1s2s^2$ and $1s2p^2$ of the resonance state interact directly with the configuration $1s^2ks$ of the ^2S scattering continuum but not with perturbing configurations $(nl)^2ks$ unless $n = 2$. New calculations are needed to explore this expected effect of target atom correlation on the resonance width.

The analysis of Andrick and Bitsch (1975) of their measured angular distributions showed that up to 19 eV the d-wave phase shift is in good agreement (approximately 5%) with the partial-wave Born approximation. For a dipole polarization potential this is

$$\tan \eta_l \cong \frac{\pi \alpha_d k^2}{(2l + 3)(2l + 1)(2l - 1)}, \qquad l > 0. \tag{6.1}$$

If Eq. (6.1) is valid for $l = 2$, it must hold for higher l values. If the further approximation is made of replacing $\tan \eta_l$ by $\sin \eta_l$, the sum of all contributions to the scattering amplitude for all phase shifts with $l > 2$ is given in closed form (Thompson, 1966). In these circumstances, Andrick and Bitsch (1975) were able to deduce the whole set of phase shifts from the observed angular dependence of the differential cross section. Since the phase shifts determine the absolute values of both the total and momentum-transfer cross sections, no external calibration is needed. The values of η_0 and η_1

inferred in this way agreed within the implied error estimates with the polarized orbital and Bethe–Goldstone values shown in Figs. 6.1 and 6.2.

At energies below 5 eV, comparison of the Bethe–Goldstone momentum-transfer cross section with experimental data indicates a discrepancy of several percent outside the experimental error estimate. The theoretical calculations did not include effects of target atom electronic correlation and did not systematically include the effect of electric quadrupole polarizability.

The first calculation to include both polarization and target atom correlation at a consistent level of accuracy was carried out by O'Malley *et al.* (1979), who used the *R*-matrix method with a correlated target wave function. Pseudostates were included to represent both dipole and quadrupole polarizabilities. Since Andrick and Bitsch had shown that phase shifts η_l for $l > 1$ could be estimated from the Born formula, only η_0 and η_1 were computed. Best estimates of η_0 and η_1 were reported, with estimated errors ranging from 0.39 to 0.25%, and from 3.23 to 0.36%, respectively, based on calculations for energies up to 16.5 eV.

In order to verify and refine these results, matrix variational calculations were carried out with a correlated helium target-state wave function (Nesbet, 1979b,c). Preliminary variational calculations on the He ground state gave a sequence of five levels of accuracy, ranging from Hartree–Fock at level 1 to a correlated function at level 5 for which the residual errors in correlation energy and dipole polarizability were 1.91 and 0.02%, respectively. At each level, a list of target atom configurations, including a $^1P^0$ variational manifold for the polarizability calculation, was specified. For matrix variational calculations of the scattering phase shifts, each target atom configuration included at a given level was augmented by adjoining all possible orbital functions in an extended orbital basis that could be coupled to target functions to give a scattering state of correct symmetry. Configurations representing quadrupole polarizability and short-range correlation were added to the target atom representation so that the indicated residual error due to these effects was less than a specified upper limit.

Phase shifts η_0 and η_1 were computed in this way for values of k from $0.1 \, a_0^{-1}$ to $1.1 \, a_0^{-1}$. Convergence with respect to level of target representation, to orbital basis, and to quadrupolar and short-range correlation effects was monitored for each k value to give phase shifts estimated by extrapolating to convergence in all three respects. The final values of phase shifts were taken to be the estimated values, with estimated error bounds given by the difference between directly computed and estimated values.

For the differential cross section to be accurate within 1%, η_0 must be computed within 0.50%. The calculations were designed so that the final estimated error of η_0 met this criterion, varying from 0.17 to 0.50% over the range of k considered. The requirement for accuracy of η_1 is less stringent than for η_0 because its contribution to the scattering amplitude is smaller throughout the energy range considered. The variational calculations maintained an estimated error bound of 0.66% or better. The net effect of the residual errors of η_0 and η_1 was no greater than 1% of the differential cross section at any scattering angle. The error in η_2 due to use of the Born formula is marginally within the same bound.

The modified effective range expansion, as applied by O'Malley (1963), would express η_0 and η_1 for small k as

$$\eta_0(k) = \tan^{-1}[-A_0(k)k(1 + \tfrac{4}{3}\alpha_d k^2 \ln k) - \tfrac{1}{3}\pi\alpha_d k^2], \qquad (6.2)$$

$$\eta_1(k) = A_1(k)k^2, \qquad (6.3)$$

where $A_0(k)$ is approximated by a quadratic function of k, and $A_1(k)$ is approximated by a linear function of k. These functional forms were used in the variational calculations to define smoothly varying functions $A_0(k)$ and $A_1(k)$. Cubic spline functions were used to interpolate these functions between values of k used for the variational calculations in the form

$$A(k) = \sum_{i=0}^{3} c_i(k_a)(k - k_a)^i, \qquad k_a \le k < k_{a+1}. \qquad (6.4)$$

Cubic spline functions are continuous at the node points k_a with continuous first and second derivatives. The coefficients c_i for $A_0(k)$ are listed in Table 6.1 and those for $A_1(k)$ in Table 6.2. In Eqs. (6.2) and (6.3), α_d is to be given its recommended experimental value, $1.384138\,a_0^3$ (Leonard and Barker, 1975). The same value is used in Eq. (6.1) for higher-order phase shifts.

Table 6.1. Coefficients for Cubic Spline Fit of Auxiliary Function $A_0(k)$ Used to Interpolate $\eta_0(k)$

$k(a_0^{-1})$	c_0	c_1	c_2	c_3
0.0	1.183511	0.000000	1.541496	−3.919410
0.1	1.195006	0.190717	0.365674	−3.919410
0.2	1.213815	0.146269	−0.810149	8.024214
0.3	1.228365	0.224966	1.597115	−12.859143
0.4	1.253973	0.158615	−2.260628	8.919655
0.5	1.256148	−0.025921	0.415268	−12.495792
0.6	1.245213	−0.317742	−3.333469	16.212961
0.7	1.196317	−0.498047	1.530419	−0.796957
0.8	1.161020	−0.215871	1.291332	−1.879405
0.9	1.150467	−0.013987	0.727510	9.953116
1.0	1.166296	0.430108	3.713445	−12.378150
1.1	1.234063	0.801453	0.000000	0.000000

In Table 6.1, the coefficient $c_0(0.0)$ is adjusted to make c_3 continuous at $k = 0.1\ a_0^{-1}$, subject to $c_1(0.0) = 0$ as required by the effective range expansion. This gives an extrapolated value of the scattering length (O'Malley, 1963)

$$A_0 = c_0(0.0) = 1.183511\ a_0, \tag{6.5}$$

with estimated error less than 0.50%. This agrees with the value $1.18\ a_0$ obtained by applying the effective range formula to the experimental momentum-transfer cross section for energies from 0.008 to 0.1 eV (Crompton *et al.*, 1967, 1970). The experimental error limit is 2%. O'Malley *et al.* (1979) give $1.177\ a_0$ as their best estimate of A_0 with estimated error $\pm 0.006\ a_0$ in agreement with the extrapolated variational value.

In Table 6.2, the coefficient $c_0(0.0)$ is given its theoretical value, in atomic units,

$$\pi\alpha_d/15 = 0.289893 \tag{6.6}$$

and $c_1(0.0)$ is adjusted so that c_3 is continuous at $k = 0.1$.

If Eq. (6.1) is valid for $\sin \pi_l$, for $l > 2$, the formula of Thompson (1966) for the scattering amplitude is

$$f(\theta) = \frac{1}{k} \sum_{l=0}^{2} (2l + 1) \exp(i\eta_l) \sin \eta_l P_l(\cos \theta)$$

$$+ \pi\alpha_d k \left[\frac{1}{3} - \frac{1}{2}\sin\frac{\theta}{2} - \sum_{l=1}^{2} \frac{1}{(2l+3)(2l-1)} P_l(\cos \theta) \right]. \tag{6.7}$$

A closed expression for the differential cross section as a smooth function of

Table 6.2. Coefficients for Cubic Spline Fit of Auxiliary Function $A_1(k)$ Used to Interpolate $\eta_1(k)$

$k(a_0^{-1})$	c_0	c_1	c_2	c_3
0.0	0.289893	0.148278	0.514098	−1.547826
0.1	0.308314	0.204663	0.049750	−1.547826
0.2	0.327730	0.168178	−0.414598	−0.122980
0.3	0.340279	0.081570	−0.451492	0.990330
0.4	0.344911	0.020981	−0.154393	−1.260623
0.5	0.344205	−0.047716	−0.532580	1.712067
0.6	0.335819	−0.102870	−0.018960	−1.302014
0.7	0.324041	−0.145732	−0.409564	0.800510
0.8	0.306173	−0.203620	−0.169411	0.475259
0.9	0.284593	−0.223244	−0.026833	0.559171
1.0	0.262559	−0.211836	0.140918	−0.469728
1.1	0.242315	−0.197744	0.000000	0.000000

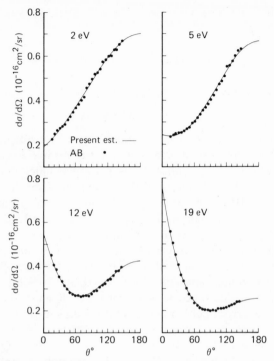

Figure 6.5. e^-–He, differential cross section (Nesbet, 1979b, Fig. 1).

θ and k is given by using this formula with η_0 and η_1 computed from Eqs. (6.2), (6.3), and (6.4) and with η_2 computed from Eq. (6.1). In atomic units a_0^2 per steradian, the differential cross section is

$$\frac{d\sigma}{d\Omega} = |f(\theta)|^2, \tag{6.8}$$

with $f(\theta)$ given by Eq. (6.7).

Absolute differential cross sections computed from these formulas (Nesbet, 1979b,c) are shown in Fig. 6.5 as smooth curves. The data points shown are the published angular distribution data of Andrick and Bitsch (AB) (1975), scaled by the ratio of the computed total cross section σ_T to their published value. The agreement is well within the experimental error bounds for all angles and energies.

The total elastic cross section obtained by integrating $d\sigma/d\Omega$ derived from Eq. (6.7) (Thompson, 1966), in units a_0^2, is

$$\sigma_T = \frac{4\pi}{k^2} \sum_{l=0}^{2} (2l+1) \sin^2 \eta_l + \frac{1}{2450}\,\pi^3 \alpha_d^2 k^2, \tag{6.9}$$

and the momentum-transfer cross section (Nesbet, 1979c), also in units a_0^2, is

$$\sigma_M = \frac{4\pi}{k^2} \sum_{l=0}^{2} (l+1) \sin^2 (\eta_l - \eta_{l+1}) + \frac{2}{33075} \pi^3 \alpha_d^2 k^2, \qquad (6.10)$$

where the first term requires the approximation

$$\sin \eta_3 = (1/315)\pi\alpha_d k^2. \qquad (6.11)$$

The second term in Eq. (6.10) is obtained by direct integration of $(1 - \cos \theta)f^2(\theta)$, using the Born approximation for terms in $f(\theta)$ with $l > 2$ then subtracting $(12\pi/k^2) \sin^2 \eta_3$ since this is already included in the first term.

The total cross section σ_T and the momentum-transfer cross section σ_M, both in units 10^{-16} cm^2, as computed from Eqs. (6.9) and (6.10) using the interpolated variationally estimated values of η_0 and η_1, are shown as full curves in Fig. 6.6. To verify the error estimates, similar calculations were carried out with the directly calculated values of η_0 and η_1, using the corresponding computed value of α_d in the Born formula. The difference between corresponding values of σ_T was less than 0.70% for all values of k considered, and the difference between corresponding values of σ_M was less than 0.66%. The error of either curve is estimated to be less than 1% for all values of k.

Directly calculated and best estimate cross sections of O'Malley et al. (1979) are shown in Fig. 6.6 (OBB). The general agreement with the variational results is within combined error limits. The variational Bethe–Goldstone calculations (Sinfailam and Nesbet, 1972) of σ_T are shown in Fig. 6.6 (SN). The new values are noticeably larger, up to 4.60% at $k = 0.1\ a_0^{-1}$. The SN values correspond to the full curve shown in Fig. 6.4. It is clear that the new theoretical values of σ_T are substantially greater than the experimental values of Golden and Bandel (1965).

Comparison with σ_T and σ_M values derived by Andrick and Bitsch (1975), shown as points (AB) in Fig. 6.6, indicates that the variational results are systematically smaller but well within the stated experimental error limits. Recent measurements of σ_T by Kauppila and collaborators (Kauppila et al., 1977; Stein et al., 1978) are shown in Fig. 6.6. The agreement is very good, compatible with a general error of 2% or less in the experimental data, although experimental error estimates were not published. Values of σ_T measured by Kennerly and Bonham (1978) are not shown in Fig. 6.6 but agree closely with the data of Kauppila et al. (1977). The variational values of σ_T shown in Fig. 6.6 are within the experimental error estimates (+3%, −2% except for $E < 2$ eV). A similar comparison is made with measurements of σ_M by Crompton et al. (1967, 1970) and by Milloy and Crompton (1977), whose data points are shown in Fig. 6.6. Here the agreement with

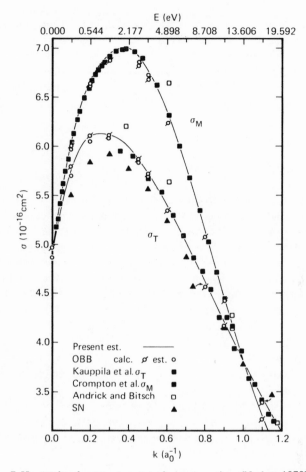

Figure 6.6. e^-–He, total and momentum-transfer cross sections (Nesbet, 1979b, Fig. 2).

the variational results is well within the combined error limits, using the experimental error estimate of $\pm2\%$ for energies in the range 0.01 to 3 eV, $\pm3\%$ in the range 4 to 7 eV, and $\pm5\%$ otherwise. In fact, if the error of the variational results is taken to be $\pm1\%$, the data of Crompton *et al.* (1967, 1970) are compatible with an error of $\pm1\%$ for the whole range of energies considered.

Comparing the variationally estimated phase shifts (Nesbet, 1979b,c) with those of O'Malley *et al.* (1979), the two sets of results for η_0 are within their combined error estimates. In the case of η_1, expressed in terms of the auxiliary function $A_1(k)$, the best-estimated values of O'Malley *et al.* fall outside the error limits of the variational results, and the difference generally exceeds the error estimated by O'Malley *et al.* The earlier variational values of η_1 (Sinfailam and Nesbet, 1972) are generally closer to the new variational values, especially at higher energies. The polarized orbital values of η_1 computed by Duxler *et al.* (1971) and the modified polarized orbital values computed by Yau *et al.* (1978) roughly bracket the new variational values, when expressed in terms of $A_1(k)$, with displacements of several percent above and below, respectively.

6.2. Helium: Inelastic Scattering

Calculations by Oberoi and Nesbet (1973c) continued the Bethe–Goldstone variational calculations of Sinfailam and Nesbet (1972) above the 2^3S threshold at 19.818 eV. For energies between the $n = 2$ and $n = 3$ energy thresholds, multichannel variational equations were solved, including all significant effects of virtual excitations of reference configuration $1s2s$ at the level of the Bethe–Goldstone hierarchy symbolized by $[2s]$. The reference configuration was augmented by $ns/2s$ and $np/2s$ virtual excitations, required for representation of the four $n = 2$ states and the 1^1S ground state. Scattering states with open-channel orbitals up to $l = 3$ were included in these calculations, and the orbital basis set was sufficiently large to ensure reasonable convergence. Continuum orbitals with $l > 3$ are not expected to make a significant contribution to structural features below the $n = 3$ threshold.

The principal results of these calculations can be discussed in terms of Fig. 6.7 (Oberoi and Nesbet, 1973c). This figure shows sums of eigenphases for the principal doublet scattering states plotted against the electron

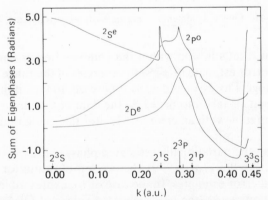

Figure 6.7. e^-–He, sums of eigenphases for $^2S^e$, $^2P^0$, and $^2D^e$ partial waves (Oberoi and Nesbet, 1973c, Fig. 1).

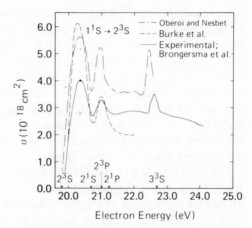

Figure 6.8. e^-–He, $1^1S \rightarrow 2^3S$ excitation cross section (Oberoi and Nesbet, 1973c, Fig. 5).

momentum k in the 2^3S channel. The main structural features in these computed curves correspond to observed features in the various elastic and inelastic cross sections coupling the five target states in this energy region.

Following the discussion in Section 3.2, a multichannel resonance corresponds to a rise through π radians of the sum of eigenphases for some symmetry component of the scattering wave function. The computed values of threshold energies are shown in Fig. 6.7. The experimental $n = 2$ thresholds are 2^3S at 19.818 eV, 2^1S at 20.614 eV, 2^3P^0 at 20.964 eV, and 2^1P^0 at 21.218 eV. From Fig. 6.7, the computed eigenphases indicate a broad $^2P^0$ resonance between the 2^3S and 2^1S thresholds, a broad ^2D resonance near the 2^3P^0 threshold, and several narrow resonances below the 3^3S threshold (22.719 eV). Unpublished calculations show that the $^2P^0$ resonance is associated with an excited He^- valence state of configuration $1s2s2p$ and the ^2D resonance with a valence state of configuration $1s2p^2$. The ^2S resonance at 19.36 eV corresponds to $1s2s^2$ perturbed by $1s2p^2$, but there is no evidence of a higher ^2S resonance associated with the orthogonal linear combination of these configurations.

The computed $1^1S \rightarrow 2^3S$ excitation cross section is shown in Fig. 6.8 in comparison with experimental data of Brongersma et al. (1972). Similar data have been obtained by Hall et al. (1972). The error bar shown in the figure indicates the large uncertainty in the absolute normalization of the experimental cross section. The three prominent peaks, in order of increasing energy, correspond to the broad $^2P^0$ and ^2D resonances and to a cluster of narrow resonances near the $n = 3$ threshold. If the experimental data were normalized to the variational calculation, the two curves would be in very close agreement. The third curve shown is from the five-state close-coupling calculation of Burke et al. (1969a). An exchange term was approximated in these calculations, and the computed s-wave partial excitation cross section was divided by a factor of 10 to reduce its contribution to the curve shown here. The five-state approximation excludes $n = 3$ states and cannot describe the resonances associated with them. The $^2P^0$ and ^2D resonances are prominent features of the observed cross section for production of the metastable states 2^3S and 2^1S (Schulz, 1973a). The p-wave character of inelastically scattered electrons at the $^2P^0$ resonance energy was identified by differential cross-section measurements (Ehrhardt and Willmann, 1967a). The same experiment established the d-wave character of scattered electrons at the ^2D peak.

Figure 6.7 shows a prominent Wigner cusp in the $^2P^0$ eigenphase sum at the 2^3P^0 threshold. A similar structure occurs at the 2^1P^0 threshold but is not visible on the scale of the figures. As shown in Section 3.3, such points of vertical slope can occur at a channel threshold in scattering states of symmetry such that the new open-channel orbital is an s wave. In the present case, cusps or rounded steps can occur in $^2P^0$ scattering states at the 2^3P^0 and 2^1P^0 thresholds and in ^2S scattering states at the 2^3S and 2^1S thresholds.

Threshold cusps show up prominently in the computed 1^1S–2^1S excitation cross section shown in Fig. 6.9 (Oberoi and Nesbet, 1973c) at the 2^3P^0 and 2^1P^0 thresholds. The rapid rise from threshold of this excitation cross section combines the effect of a virtual state, in the ^2S scattering state (Burke *et al.*, 1969a) with the $^2P^0$ and ^2D resonances. Resonance structure appears

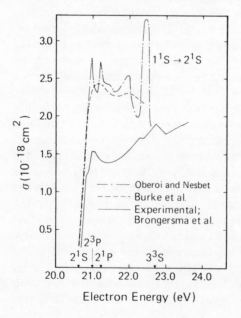

Figure 6.9. e^-–He, $1^1S \rightarrow 2^1S$ excitation cross section (Oberoi and Nesbet, 1973c, Fig. 7).

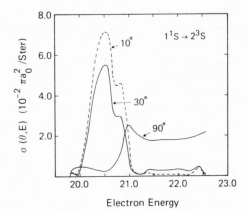

Figure   6.10. e^-–He, $1^1S \to 2^3S$ differential cross section (Oberoi and Nesbet, 1973c, Fig. 6).

below the $n = 3$ threshold. The close-coupling calculation (Burke *et al.*, 1969a) is also shown in the figure and would be in substantial agreement with the variational curve if threshold effects were smoothed out and the $n = 3$ resonances were omitted. Comparison with the experimental data of Brongersma *et al.* (1972) is less satisfactory for the 2^1S excitation than for the 2^3S excitation. Structure appears in the experimental curve shown in Fig. 6.9, in the region of the 2^3P^0 and 2^1P^0 thresholds and of the $n = 3$ resonances, but the general increase of the experimental cross section with energy is not reflected in the theoretical cross section. Improved five-state calculations, using the R-matrix method (Berrington *et al.*, 1975b), do show this general increase with energy, due in part to inclusion of $^2F^0$ and ^2G scattering states in the calculations. Oberoi and Nesbet (1973c) included only ^2S, $^2P^0$, and ^2D states.

The smooth descent of the ^2S eigenphase sum from the 2^3S threshold, shown in Fig. 6.7, is characteristic of threshold behavior in elastic scattering when a true bound state lies just below the threshold, as indicated in Eq. (3.92). In the present case, two channels are open, but the narrow ^2S resonance near 19.36 eV has the same analytic effect on the 2^3S channel opening at 19.818 eV as a similarly displaced bound state would have on a single elastic scattering channel. Ehrhardt *et al.* (1968) give a detailed argument based on Eq. (3.92) and on the assumption of weak coupling between the two ^2S eigenchannels near threshold to account for s-wave structure found at the 2^3S threshold in their experimental differential excitation cross section. Computed differential excitation cross sections are shown as functions of energy in Fig. 6.10 (Oberoi and Nesbet, 1973c). The structure considered by Ehrhardt *et al.* appears as a peak at threshold at all angles. These calculations confirm the analysis of Ehrhardt *et al.*, who showed that the s-wave threshold structure is a consequence of the ^2S

resonance below threshold, influencing the new eigenchannel phase shift through Eq. (3.92). Since the width of the 19.36-eV resonance is only 0.01 eV, the phase shift has completed its rise through π radians at the threshold and has no direct influence there. Thus the *width* of the 19.36-eV resonance is irrelevant to the observed structure, which depends primarily on the parameter β in Eq. (3.92), defined by the energy *displacement* of the resonance below the threshold. This contradicts a qualitative argument given by Taylor (1970), who attributed this structure to an effect of the tail of the ^2S resonance extending beyond the 2^3S threshold.

Differential excitation cross sections in the $n = 2$ region have been computed in the five-state close-coupling approximation by Wichmann and Heiss (1974), who used the algebraic close-coupling method, and by Sinfailam (1976) and Fon et al. (1978), who used the R-matrix method. All of these results are in good agreement with the variational results shown in Fig. 6.10 and with experimental data. The most detailed results are those of Fon et al. (1978), who compare calculated excitation cross sections with absolute cross-section data of Pichou et al. (1976). In general, theory and experiment show the same principal structural details, and the agreement is essentially quantitative for all scattering angles at energies below 21 eV. Some quantitative effect of the $n = 3$ states may be missing above this energy.

The R-matrix calculations of Sinfailam (1976) were carried out with and without the ^2D component of the scattering wave function. Comparison of these results showed that the peak near threshold in the $90°\ 2^3S$ excitation cross section, shown in Fig. 6.10, is due to the ^2S component of the wave function, but the subsequent dip results from interference with the ^2D component, which varies rapidly due to the ^2D resonance.

The very sharp rise of the ^2S eigenphase sum at the 2^1S threshold, shown in Fig. 6.7, is an example of threshold behavior characteristic of a virtual state. From Eq. (3.90), $\tan \eta_v$ rises linearly with k, but the phase shift cannot rise more than $\pi/2$ radians. A slope of β^{-1} at threshold results from a virtual state on the nonphysical energy sheet, $-\beta^2/2$ below the threshold. Prominent cross-section structure occurs only if β is quite small. Observed structure near the 2^1S threshold was associated with a virtual state by Ehrhardt et al. (1968). The virtual state was first identified in close-coupling calculations (Burke et al., 1969a).

Detailed variational calculations have been carried out for energies close to the 2^3S and 2^1S thresholds, where cusp structure can occur in the ^2S scattering state (Nesbet, 1975b). At the 2^3S threshold, a rounded step structure occurs in the computed total elastic cross section, in quantitative agreement with parameters used to fit the threshold structure in the observed differential cross section (Cvejanovic et al., 1974).

The experimental data at the 2^3S threshold were fitted (Cvejanovic et al., 1974) to the linear two-channel formulas given here by Eqs. (3.121), (3.122), and (3.123), with parameter values

$$\eta = 1.87 \pm 0.15 \text{ radians},$$
$$b = 2\chi^2 \sin^2 \eta = (13 \pm 3) \times 10^{-10} \text{ cm}. \tag{6.12}$$

In Fig. 6.11 (Nesbet, 1975b), the curves denoted by σ_{exp}, $\underline{\sigma}_{exp}$, and $\bar{\sigma}_{exp}$ indicate the mean and extremal values of the elastic cross section σ_{11} computed from the linear two-channel formulas, using the experimental parameters, for the ^2S partial cross section. Theoretical values of the $^2P^0$ and ^2D partial cross sections have been added in, and all curves are adjusted to the best-estimated value $\eta(^2S) = 1.8278$ at the threshold. The curve σ_{theory} is obtained by fitting the variationally computed K matrices above the 2^3S threshold with nonlinear formulas and then using the multichannel threshold theory of Section 3.3 to extrapolate below threshold, retaining linear terms only.

Figure 6.11 indicates excellent agreement between experiment and theory within the limits of the linear threshold formulas, which are used below threshold for all curves shown. Above threshold, the theoretical cross section rises above the linear fit to the experimental threshold data because it contains terms nonlinear in k. The close agreement below threshold

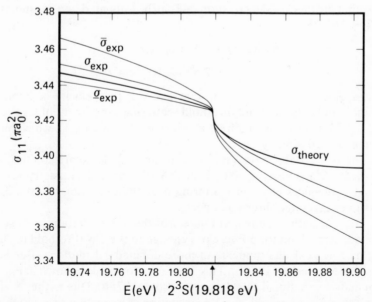

Figure 6.11. e^-–He, total elastic cross section at the 2^3S threshold (Nesbet, 1975b, Fig. 2).

checks the validity of the theory used to extrapolate computed data from above to below a threshold. The linear parameter values obtained by fitting the variational calculations were (Nesbet, 1975b)

$$\eta = 1.8278 \text{ radians,}$$
$$b = 0.1968, \qquad a_0 = 10.4 \times 10^{-10} \text{ cm,} \tag{6.13}$$

within the experimental error bounds indicated in Eqs. (6.12).

From Eq. (3.123), the parameter b determines the rate of increase from threshold of the excitation cross section. The variational calculations show that this remains nearly linear in k_2 (with respect to the 2^3S channel), including both ^2S and $^2P^0$ partial cross sections, up to $k_2 = 0.10 \, a_0^{-1}$. Hence the corresponding transition rate (cross section times k_2) for electron-impact de-excitation $2^3S \to 1^1S$ is nearly constant in this range since

$$k_2\sigma_{21} = (k_1^2/3k_2)\sigma_{12} = (2\pi/3)b. \tag{6.14}$$

The value implied by Eq. (6.13) is

$$k_2\sigma_{21} = 0.1312 \, \pi a_0, \qquad k_2 \le 0.10 \, a_0^{-1}. \tag{6.15}$$

Nesbet et al. (1974) showed that the rate coefficient $K(2, 1)$ for deactivation of metastable He(2^3S) by thermal electrons can be estimated from $k_2\sigma_{21}$ by equating the temperature in degrees Kelvin to $k_2^2/3k_B$, where k_B is the Boltzmann constant, 3.1667×10^{-6} a.u./°K. This argument and Eq. (6.15) imply that the thermal deactivation rate can be estimated by a constant value for $T \le 1053$°K,

$$K_{\text{He}}(2, 1) = 0.1312(\pi \alpha c a_0^2)$$
$$= 2.525 \times 10^{-9} \text{ cm}^3/\text{sec.} \tag{6.16}$$

Here α is the fine-structure constant and c is the velocity of light. This argument indicates that the threshold scattering experiment of Cvejanovic et al. (1974) provides an indirect measurement of $K_{\text{He}}(2, 1)$, which has not been measured directly.

The $n = 2$ states have very large electric dipole polarizabilities: $\alpha_d = 316 \, a_0^3$ for 2^3S and $\alpha_d = 802 \, a_0^3$ for 2^1S (Chung and Hurst, 1966). This strongly influences electron scattering near the excitation thresholds and leads to exaggerated threshold effects.

The calculations shown in Fig. 6.9 of the 2^1S excitation cross section were not carried out for points close enough to the 2^1S threshold to show a very narrow peak that appears in more detailed calculations (Berrington et al., 1975b; Nesbet, 1975b). This 2^1S excitation peak is shown in Fig. 6.12 from matrix variational calculations (Nesbet, 1975b). Due to the ^2S virtual state at the 2^1S threshold, the 2^1S excitation cross section rises much more

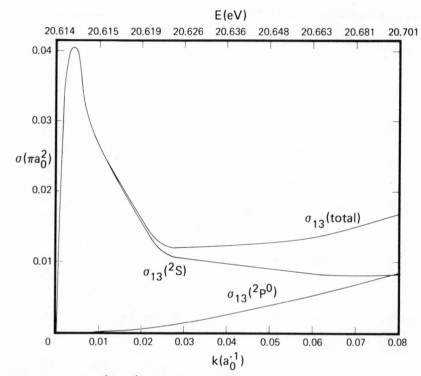

Figure 6.12. e^-–He, $1^1S \to 2^1S$ excitation cross section near threshold (Nesbet, 1975b, Fig. 5).

rapidly from threshold than does the 2^3S cross section. This agrees with high-resolution experiments on helium threshold excitation (Cvejanovic and Read, 1974), which indicate that the 2^1S threshold excitation peak as measured becomes progressively larger than the corresponding 2^3S peak as experimental energy resolution is improved.

Figure 6.13 shows the computed 2^1S and 2^3S excitation cross sections on a common energy scale (Nesbet, 1975b). The ratio $R(\Delta E)$ of the integrals of these cross sections up to ΔE is shown as a function of energy ΔE above either threshold. This ratio can be compared to the relative signal measured in trapped-electron experiments. Cvejanovic and Read (1974) find that the 2^1S to 2^3S threshold peak ratio, which corresponds to $R(\Delta E)$ for an experimental acceptance width ΔE, increases with experimental energy resolution. At the best attainable resolution with acceptance width 16 meV (width at half-height of the extraction efficiency curve), the experimental peak ratio is 2.6. From Fig. 6.13, the computed ratio R for this value of ΔE is 2.5, in quantitative agreement with experiment. This ratio drops to values near unity for ΔE in the range 50–100 meV, in agreement with earlier

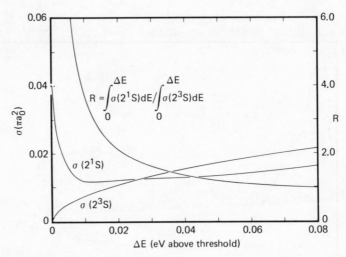

Figure 6.13. e^-–He, 2^1S and 2^3S excitation cross sections and ratio of their energy integrals (Nesbet, 1975b, Fig. 6).

experiments of lower-energy resolution. By implication, the computed function $R(\Delta E)$ can be used to calibrate the energy resolution of trapped-electron experiments.

The multichannel threshold theory of Section 3.3 was used (Nesbet, 1975b) to compute cusp and rounded-step structures for all of the background elastic and inelastic cross sections at both the 2^3S and 2^1S thresholds. The inelastic background cross sections at the 2^1S threshold have not been observed in sufficient detail to provide data for comparison with the theory.

Because $n = 3$ states were not represented explicitly in the calculations of Oberoi and Nesbet (1973c), the details of resonance structures associated with these states were not given accurately. More refined variational calculations were carried out (Nesbet, 1978a) that included explicit $n = 3$ basis orbitals. Continuum Bethe–Goldstone equations were solved variationally, using the OAF method (Nesbet and Oberoi, 1972), for two electrons outside a He^+ core. This corresponds to the $[2s]$ level of approximation, with respect to reference configuration $1s2s$. The basis set used eight s, seven p, four d, and four f orbitals plus continuum functions with orbital $l \leq 3$ for all energetically allowed open channels. Total symmetry states ^2S, $^2P^0$, and ^2D were considered.

The computed total cross section for excitation from 1^1S to the 2^3S and 2^1S metastable states is shown in Fig. 6.14. In the figure, k_2 is the wave vector relative to the 2^3S threshold, defining all energies in the variational calculations. The $n = 3$ threshold values, calculated as excitation energies relative to the experimental 2^3S energy, are shown in Table 6.3. They are in good agreement with spectroscopic values (Martin, 1973). The displacements are of sufficient size, however, that the computed and experimental energy scales cannot be superimposed exactly. Since $^2F^0$ and higher scattering states are omitted, a background rising with energy but not expected to have sharp structural features should be added to the theoretical cross section.

Brunt *et al.* (1977) have measured the cross section for electron-impact excitation of He metastable states. Their data, shown in Fig. 6.15, resolves structural details at an energy resolution (15 meV) not previously achieved. Comparison of Figs. 6.14 and 6.15 indicates close correspondence between theory and experiment for all structural features resolved by the experiment. The theoretical calculations also show fine details at the 3^3S and 3^1S thresholds that are not apparent in the experimental data. The 3^3S threshold structure contains inverted cusps in both the 2^3S and 2^1S excitation cross

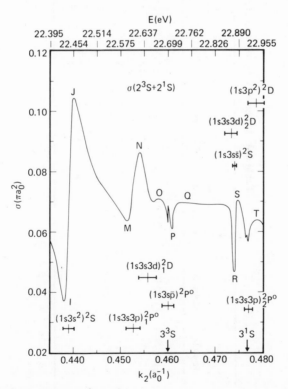

Figure 6.14. e^-–He, computed $2^3S + 2^1S$ excitation cross section (Nesbet, 1978a, Fig. 1).

Table 6.3. He energy Levels and He⁻ Resonances

He state	E (eV)		Feature[b]	He⁻ state	E (eV)		Γ (meV)	
	Observed[a]	Calculated			Observed[b]	Calculated	Observed[b]	Calculated
$1s3s\,^1S$	22.921	22.914	T	$1s3p^2\,^2D$	22.93	22.938	20	43.3
			S	$1s3s3p\,^2P^0$	22.89	22.913		18.7
			RS	$1s3s\bar{5}\,^2S$	22.88	22.877	18	6.7
			RS	$1s3s3d\,^2D$	22.88	22.867		30.4
			PQ	$1s3s\bar{p}\,^2P^0$	22.74	22.703		23.3
$1s3s\,^3S$	22.719	22.698	P					
			NO	$1s3s3d\,^2D$	22.66	22.645	20	44.7
			MN	$1s3s3p\,^2P^0$	22.62	22.608	38	34.9
			IJ	$1s3s^2\,^2S$	22.45	22.441	36	25.2

[a] Martin (1973).
[b] Brunt *et al.* (1977).

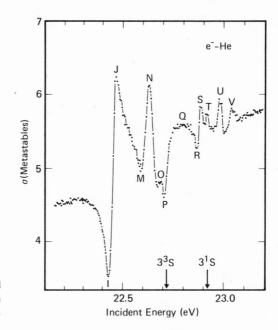

Figure 6.15. e^-–He, observed metastable excitation cross section (adapted from Brunt *et al.*, 1977).

sections. This is accompanied by the very rapid initial rise of the 3^3S excitation cross section shown as $\Sigma\delta$ in Fig. 6.16, peaking at 0.2 meV above threshold. The sum of eigenphases in the ^2S scattering state rises rapidly from the threshold, as shown in Fig. 6.16, which is characteristic of a virtual state. The 3^1S threshold structure contains rounded steps in both 2^3S and 2^1S excitation cross sections, accompanied by a narrow ^2S resonance just below threshold and a rapid initial rise in the 3^1S excitation cross section, as shown in Fig. 6.17. The rapid *decrease* from threshold of the sum of ^2S eigenphases, shown in Fig. 6.17 as $\Sigma\delta$, is characteristic of a resonance or bound state just below threshold.

A systematic search for resonances was carried out in the theoretical calculations using methods described in Sections 3.2 and 3.4. Following ideas of the stabilization method (Taylor, 1970) an eigenvector of the bound–bound matrix, which treats He$^-$ as a bound atomic system, may correspond to a resonance if it is dominated by a particular state constructed from ground or excited valence-shell orbitals. All eigenvalues in the energy range of interest ($0.43 < k_2 < 0.48$) were examined by the resonance search procedure. The results, listed in Table 6.3, for state designations, resonance energies, and widths, can be associated with the labels used by Brunt *et al.* (1977), as shown in Fig. 6.15, to designate observed structural features. The computed resonance positions and widths are shown graphically in

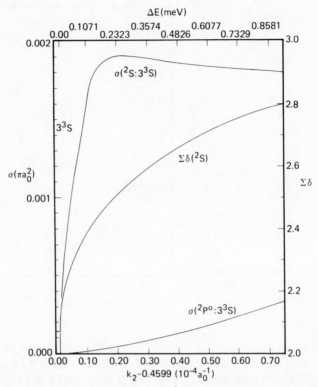

Figure 6.16. e^-–He, structure near the 3^3S threshold (Nesbet, 1978a, Fig. 3).

Fig. 6.14 by small vertical and horizontal bars, respectively, for each computed E_r and Γ_r.

The general agreement between theory and experiment is very good. In the particular case of the ^2D resonance labeled NO, it has been shown by Andrick (1979), from measurements of the 2^3S differential excitation cross section, that the width is 43–50 meV, rather than 20 meV as given by Brunt *et al.* (1977). Andrick's results for the widths of the first three resonances (^2S, $^2P^0$, and ^2D) are 30–35 meV, 37–45 meV, and 42–50 meV, respectively, in good agreement with the calculated widths listed in Table 6.3.

Two classes of resonance states were found. For the first, exemplified by $(1s3s^2)^2S$, the specified state is the largest single bound-state component of the wave function, and the resonance is associated with the particular eigenvector that contains the largest component of the designated state. These resonances are excited valence shell (VS) states of He$^-$ in the $n = 3$ shell. This correspondence fails for the second class of resonance states,

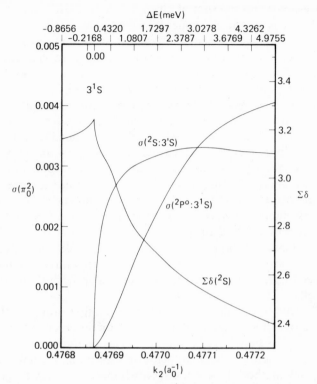

Figure 6.17. e^-–He, structure near the 3^1S threshold (Nesbet, 1978a, Fig. 4).

$(1s3s\bar{p})^2P^0$ and $(1s3s\bar{s})^2S$, close to excitation thresholds. Since they are *not* uniquely associated with $n = 3$ valence-shell configurations, these states are presumably formed by attachment of an electron in the strong polarization potential of the threshold state. They correspond to nonvalence (NV) resonance states.

The structures found at the 3^3S and 3^1S thresholds appear to provide a realization of the textbook distinction between bound and virtual states. At the 3^3S threshold, a virtual state occurs but no $1s3s\bar{s}$ resonance, while the 3^1S threshold provides the opposite case. This indicates that the effective 3^3S potential is not quite strong enough to bind an s-wave electron, so the S matrix has a virtual state pole on the negative imaginary axis of k_6 (wave vector relative to the 3^3S threshold). The sum of eigenphases increases from threshold. At the 3^1S threshold, the effective potential is strong enough to bind an s state, producing the narrow $1s3s\bar{s}$ resonance below threshold. The S-matrix pole has moved to the positive imaginary axis of k_7 (relative to the 3^1S threshold) and the sum of eigenphases decreases from threshold. The virtual state is absent, but appears instead as an NV resonance.

Beyond $n = 3$, the number of target atom states increases rapidly, making it very difficult to carry out calculations without artificial constraints. In the intermediate energy region, above the ionization threshold of He at 24.5 eV, a continuous number of two-electron channels are open.

Thomas and Nesbet (1974) reported preliminary results of matrix variational calculations of the 1^1S–2^3S excitation cross section at 29.6 eV. The variational wave function represented the 1^1S ground state and all four $n = 2$ states as open channels, with virtual excitation structure equivalent to the Bethe–Goldstone approximation with reference configuration $1s^2$. Other open channels were suppressed simply by omitting the corresponding open-channel orbitals from the variational wave function. The computed differential excitation cross section is shown as the curve labeled CI in Fig. 6.18. It is in reasonable agreement with experimental data of Trajmar (1973). In particular, it shows a sharp dip at 125°, evident in experimental data. This structure had not been obtained in any previous calculation. This is exemplified by the curve labeled RPA, obtained by the first-order Green's function method (Thomas et al., 1974a).

These variational calculations differ from first-order methods by including the dynamical polarization effect of the interaction between 2^3S and 2^3P^0 as well as between 2^1S and 2^1P^0. This multichannel effect is not included in methods that include explicitly only the initial and final states of a transition. Such methods have been successful in this energy range for the dipole-allowed 2^1P^0 excitation.

The calculation of the 2^3S excitation cross section was successful only because pseudostate resonances did not occur near 29.6 eV. An attempt to apply the same method to the 2^1P^0 excitation failed for this reason. Many overlapping pseudostate resonances were present, precluding a meaningful calculation unless some systematic procedure can be found for removing the resonances or averaging over them.

6.3. Carbon, Nitrogen, and Oxygen

Experimental measurements of electron scattering by these atoms require molecular dissociation. The few experiments that have been carried out at low energies are reviewed by Bederson and Kieffer (1971). The only available results of reasonably high precision are observations of narrow resonance structures in the $n = 3$ excitation range of oxygen. The general trend of total ground-state cross sections has been measured for e^-–O

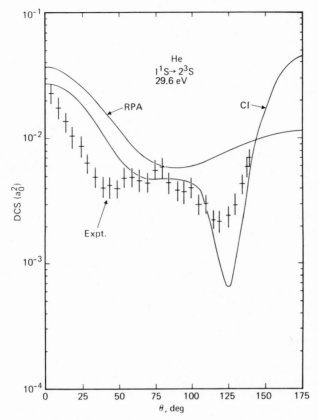

Figure 6.18. e^-–He, $1^1S \to 2^3S$ differential cross section at 29.6 eV (Thomas and Nesbet, 1974, Fig. 1).

scattering (Neynaber *et al.*, 1961; Sunshine *et al.*, 1967) and for e^-–N scattering (Neynaber *et al.*, 1963; Miller *et al.*, 1970). The ratio of forward-to-backward e^-–O scattering from 3 to 20 eV and the total differential cross section at 5 and 15 eV have been measured by Dehmel *et al.* (1974, 1976). There are no comparable data for e^-–C scattering.

C, N, and O have been several distinct states in the low-energy range because of the open-shell structure of their ground-state configurations, $2s^2 2p^n$. The ^3P ground states of C and O have static quadrupole moments that couple partial waves l and $l \pm 2$. Electric dipole polarizabilities are small compared to the alkali metal atoms, but polarization potentials still dominate low-energy scattering.

Oxygen

The polarized orbital method (Temkin, 1957) was applied to e^-–O scattering, including only the $l = 0$ partial wave and the $2p \to d$ part of the electric dipole polarizability. This work was completed, including $2p \to s$ and $2s \to p$ polarization effects, by Henry (1967). The total elastic cross section computed by Henry is shown in Fig. 6.19, in comparison with other theoretical calculations and with experimental cross-section data (Sunshine *et al.*, 1967). The experimental points, with estimated error of 20%, do not define a smooth curve. The very low values of the cross section near threshold computed by Henry are in agreement with the value deduced from shock tube measurements, 2×10^{-16} cm² (2.3 πa_0^2) at 0.5 eV (Lin and Kivel, 1959).

Close-coupling calculations, including all states of the ground configuration, were first carried out for e^-–O scattering by Smith *et al.* (1967). An algebraic error in this work was corrected in subsequent calculations by Henry *et al.* (1969). These results agree closely with a matrix variational calculation, in the single configuration (SC) approximation, by Thomas *et al.* (1974b). This computed cross section is shown in Fig. 6.19. The single configuration approximation neglects the atomic polarizability. The resulting cross section is much too large at threshold and apparently too large for energies up to 6 eV.

The curve labeled CI in Fig. 6.19 is a matrix variational calculation that includes effects due to the near degeneracy of the $2s$ and $2p$ orbitals (Thomas, 1974b). In these calculations, configurations $2s^2 2p^4$, $2s 2p^5$, and

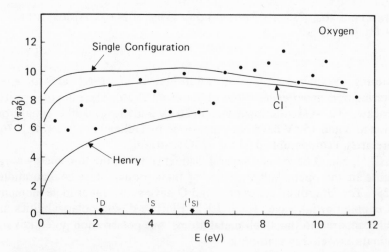

Figure 6.19. e^-–O, total cross section, early results (Thomas *et al.*, 1974b, Fig. 5).

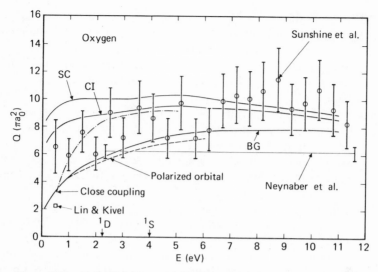

Figure 6.20. e^-–O, total cross section (Thomas and Nesbet, 1975a, Fig. 1).

$2p^6$ were included in the target-state variational basis. The second of these contributes to the dipole polarizability, and the third displaces the ^1S threshold relative to the other atomic states (^3P and ^1D). In Fig. 6.19, (^1S) denotes the computed threshold (relative to ^3P) in the single configuration approximation, while ^1S denotes this threshold in the CI approximation. The CI total cross section is uniformly smaller than the SC curve but is still much greater than the polarized orbital cross section of Henry. Neither theory nor experiment appears to give conclusive results.

This situation was clarified by subsequent calculations. Figure 6.20 shows the previous results, including 20% error bars on the experimental points. The total cross section obtained by the matrix variational method in the Bethe–Goldstone approximation is shown as the curve labeled BG (Thomas and Nesbet, 1975a,b). Effects of $2s \rightarrow np$ virtual excitations, beyond $n = 2$, were found to be small, and the calculations were simplified by omitting this class of virtual excitations. Basis orbitals and partial waves with $l \le 3$ were included in the calculations. Virtual polarization effects due to $2s \rightarrow 2p$, $2p \rightarrow ns$, $2p \rightarrow nd$ were included in the BG approximation for all three states of the oxygen ground configuration as well as short-range correlation between the external electron and the outer shell of the target atom. As shown in Fig. 6.20, the remarkable effect of this systematic inclusion of electron pair correlation and polarization is to bring the BG cross section into close agreement with the polarized orbital calculation of Henry (1967).

The shock-tube result of Lin and Kivel (1959) and the least-square line used by Neynaber *et al.* (1961) to represent their data, are included in Fig. 6.20, together with the experimental data of Sunshine *et al.* (1967). The BG results are consistent with all these data. An additional experimental test is shown in Fig. 6.21, which compares computed and observed values of the ratio of forward-to-backward scattering (Dehmel *et al.*, 1974). The BG results lie within the experimental error bars. The forward/backward ratio differs from unity at zero energy because of the electric quadrupole potential of the ^3P oxygen ground state.

Several more recent close-coupling calculations of e^-–O scattering have augmented the single configuration wave function with correlation terms and polarization pseudostates (Saraph, 1973; Rountree *et al.*, 1974; Tambe and Henry, 1976a,b). The computed total elastic cross sections are shown in Fig. 6.22. Curves SC and BG are labeled as in Fig. 6.20. An error in the SC curve has been corrected. Curve R (Rountree *et al.*, 1974) is labeled "close coupling" in Fig. 6.20. Curve S refers to Saraph (1973). PS refers to polarization pseudostate calculations of Tambe and Henry (1976a,b), which included pseudostates constructed from $\bar{s}$ and $\bar{d}$ polarization functions, but omitted specific short-range correlation terms. The effect of polarization function $\bar{p}$ was found to be small, so it was omitted, as in the BG calculations. In general, the various close-coupling results fall in between the CI and BG curves shown in Fig. 6.20.

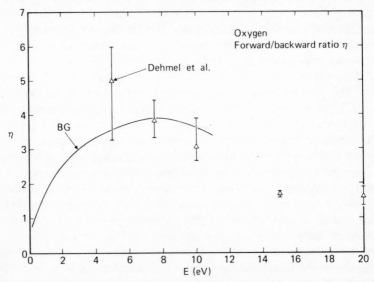

Figure 6.21. e^-–O, ratio η of forward-to-backward scattering (Thomas and Nesbet, 1975a, Fig. 5).

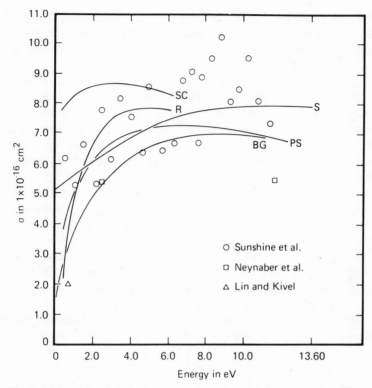

Figure 6.22. e^-–O, total elastic cross section (Tambe and Henry, 1976a, Fig. 2).

R-matrix calculations (LeDourneuf *et al.*, 1975; LeDourneuf, 1976) in the framework of the "polarized frozen core" approximation (LeDourneuf *et al.*, 1976) are structurally equivalent to the work of Tambe and Henry (1976a,b) but include $2s \rightarrow 2p$ effects and $\bar{s}$, $\bar{p}$, and $\bar{d}$ polarization functions. The e^-–O cross sections computed below 6 eV are quite similar. Both results lie above the BG cross section down to the elastic threshold.

An important approximation in the polarized pseudostate calculation of Tambe and Henry (1976a,b), which makes their theoretical model structurally different from the BG calculation, is that each polarization pseudostate is retained as a unit in the close-coupling expansion, defining a single closed-channel state. Thus only three pseudostate channels are included in the calculations, one each for $^3S^0$, $^3P^0$, and $^3D^0$, even though these states have several different components expressed as states of O^+ coupled to polarization functions. In the BG expansion, each component of the polarization pseudostate is coupled to all available orbital functions, and the coefficient of each resulting function in the $(N + 1)$-electron Hilbert space basis is determined independently. The matrix variational method, as used in the BG approximation, involves complete uncoupling of the closed-channel states. This is beyond the present capability of the close-coupling method or its R-matrix equivalent.

Comparison of the calculations of Rountree *et al.* (1974), labeled R in Fig. 6.22, with those of Tambe and Henry (1976a,b), labeled PS, indicates that uncoupling the pseudostate functions of given symmetry into separate closed-channel terms tends to reduce the computed low-energy cross section. The calculations of Rountree *et al.* treated individual polarization components as separate closed-channel terms, but polarization terms were included only for the $l = 0$ partial wave states.

The BG results are affected by an imbalance between electronic correlation energy computed for the target atom and for negative ion or electron scattering states. The consequences of this imbalance were examined for the $^2P^0$ component of the e^-–O scattering wave function (Thomas *et al.*, 1974b). The residual difference of correlation energies between the $O^-(^2P^0)$ state of configuration $2s^1 2p^5$ and the $O(^3P)$ target atom state was parameterized. Variation of this parameter by ±0.5 eV had an insignificant effect on the elastic scattering cross section.

A significant difference between the BG and PS calculations appears in the differential cross section, shown at 5 eV in Fig. 6.23 (Tambe and Henry, 1976b). The additional curve DW denotes a polarized-orbital distorted wave calculation by Blaha and Davis (1975). Since the DW calculation

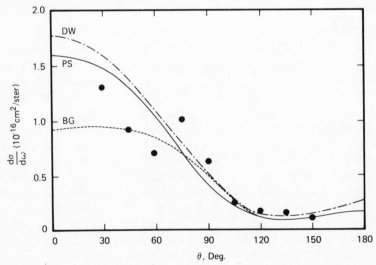

Figure 6.23. e^-–O, differential cross section at 5 eV (Tambe and Henry, 1976b, Fig. 1).

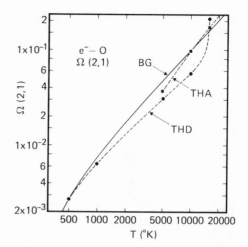

Figure 6.24. e^-–O, $^3P_2$–$^3P_1$ collision strength (LeDourneuf and Nesbet, 1976, Fig. 1).

cannot describe short-range correlation effects, but does include the full ground-state polarizability as an *ad hoc* polarization potential, it provides a model calculation with physical content similar to the pseudostate approximation (PS here). The BG differential cross section falls below the experimental point at 30°, although, as shown in Fig. 6.21, the integrated forward/backward ratio falls within the experimental error bars. The apparent structure in the experimental differential cross section between 30° and 90° is not found in any of the computed curves. More precise experimental data are needed at low impact energies in order to make a conclusive choice among the theoretical results.

The K matrix for e^-–O scattering computed by Thomas and Nesbet (1975a) was used by LeDourneuf and Nesbet (1976) to compute collision strengths for electron-impact fine-structure transitions ^3P_J–$^3P_{J'}$. The K matrix, computed in LS coupling, was transformed to jj coupling (Saraph, 1972), and the resulting excitation cross sections were shifted to energy thresholds defined by the fine-structure energy levels. The collision strength for transition $p \leftrightarrow q$ is defined as

$$\Omega_{pq} = k_p^2 \omega_p \sigma_{pq} = k_q^2 \omega_q \sigma_{qp}, \qquad (6.17)$$

where ω is the degeneracy factor of the initial state of the transition.

Figure 6.24 shows the computed collision strength (BG) for the $^3P_2$–$^3P_1$ transition as a function of electron temperature. The BG result is compared with a full jj-coupling calculation (THD) that used a simplified close-coupling wave function (Tambe and Henry, 1974) and with the result (THA) of transforming the PS calculation from LS to jj coupling, without shifting the fine-structure thresholds (Tambe and Henry, 1976a). There is

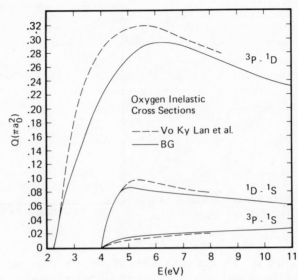

Figure 6.25. e^-–O, comparison of excitation cross section (Thomas and Nesbet, 1975a, Fig. 7).

substantial agreement between the BG curve for $\Omega(2, 1)$ and the two results of Tambe and Henry in their respective ranges of validity. Formulas fitted to this BG curve and also to the results of Tambe and Henry have been used by Hoegy (1976) to recompute the thermal rate constant for electron cooling by excitation of fine-structure levels of atomic oxygen, an important process in the dynamics of the earth's ionosphere. The new cooling rate is significantly smaller than that previously used in upper-atmosphere models.

Cross sections for excitation of the ^1D and ^1S states of the oxygen ground configuration and for scattering from these states have been computed by Thomas and Nesbet (1975a,b) in BG calculations up to 10 eV and by LeDourneuf *et al.* (1975) and LeDourneuf (1976) in R-matrix polarized frozen core calculations up to 45 eV. The BG calculations included differential cross sections for excitation of ^1D and ^1S from the ^3P ground state. The two sets of theoretical results are in general agreement, but there are no experimental data for comparison. Figure 6.25 shows the BG excitation cross sections (full curves) compared with earlier calculations by Vo Ky Lan *et al.* (1972) (dashed curves), who augmented the single-configuration close-coupling expansion with a multichannel polarized orbital wave function. The agreement is excellent for all three cross sections.

Nitrogen

Figure 6.26 shows the e^-–N total cross section computated by several different methods. Matrix variational SC and CI calculations (Thomas *et al.*, 1974b) are compared with a multiconfiguration close-coupling (MCC) calculation (Ormonde *et al.*, 1973), with a polarized orbital calculation by Henry (1968) and with experimental data of Neynaber *et al.* (1963). As in the case of oxygen, the CI calculation takes into account the $2s$, $2p$ near degeneracy by including configurations $2s^2 2p^3$, $2s 2p^4$, and $2p^5$ for the target atom. The MCC calculation is similar in structure but includes only the ^4P term of the configuration $1s^2 2s 2p^4$. Terms from configurations $2s^2 2p^2 3s$ and $2s^2 2p^2 3d$ are also included. A ^3P resonance, corresponding to the ground state of N^-, with configuration $2s^2 2p^4$, appears at 1 eV in the SC calculation but moves down to or below the scattering threshold in the CI calculation. The polarized orbital calculation gives no information about a low-energy resonance but indicates that long-range polarization effects will lower the CI or MCC cross sections. All calculations lie above the experimental error bars.

The $2s^2 2p^3$ ground configuration of N^- has states ^3P, ^1D, and ^1S, none of which have been clearly identified experimentally. Electron affinities of open-shell atoms including nitrogen were computed by Moser and Nesbet

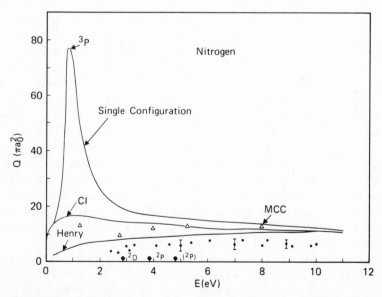

Figure 6.26. e^-–N, total cross section, early results (Thomas *et al.*, 1974b, Fig. 3).

(1971) using a variational formalism that included three-electron correlation effects. The computed energies for C^- and O^-, relative to ground states of the neutral atoms, were -1.29 and -1.43 eV, respectively, in good agreement with observed values. The computed energy of $N^-(^3P)$ was 0.12 eV, which would be a resonance energy in the scattering continuum. The ^3P state of N^- is indicated to be either very weakly bound or a low-lying electron scattering resonance, depending on very small differences of correlation energy contributions to the negative ion and neutral ground states. This is compatible with the CI calculation of Thomas et al. (1974b) but indicates that the resonance found near 1 eV in SC calculations is qualitatively incorrect.

Similar conclusions were reached on the basis of R-matrix calculations that included polarization pseudostates in the close-coupling expansion (Burke et al., 1974). The ^3P resonance moved very close to threshold when polarization and correlation terms were added to the single-configuration wave function. A six-state calculation gave the resonance position as 0.06 eV. Polarized frozen-core calculations, using the R-matrix method (LeDourneuf et al., 1975; LeDourneuf et al., 1976; LeDourneuf, 1976), place the ^3P state of N^- at 0.057 eV in the approximation used for scattering calculations and at -0.004 eV if two-electron correlation terms are included both for core states and for polarization pseudostates.

The bound-state calculations of Moser and Nesbet (1971) showed that three-electron correlation energy differences can significantly influence the computed electron affinities of complex atoms. Since, in any existing formalism, the inclusion of such terms in the scattering problem is impractical, any *ab initio* calculation of the energies of negative ions relative to neutral states will contain a residual error. A parameter Δ can be introduced as an adjustable correction for this residual net correlation energy difference (Nesbet, 1973). Calculations in the CI approximation (Thomas et al., 1974b) showed that this parameter, used to bias the energy mean values of basis states constructed from the ground-state configuration of the negative ion, could be varied over a rather wide range with internally consistent results. Energies of resonances were found to vary linearly with Δ, and the widths approached zero smoothly as the resonances approached the scattering threshold. This indicates that Δ can be adjusted to match computed resonance energies to their experimental values or to determine resonance energies by fitting observed cross sections (Nesbet, 1973).

Matrix variational calculations of e^-–N scattering, in the Bethe–Goldstone (BG) approximation, were carried out by Thomas and Nesbet (1975c,e). Several values of the parameter Δ were used. The level of virtual excitation was $[2s2p]_1$, as defined by Eq. (1.64) and by the discussion given in Section 1.6. This means that all possible single virtual excitations of target-state orbitals $2s$ or $2p$ were included in the bound component of the scattering wave function. If the orbital basis set is complete, this level of virtual excitation includes the full first-order effect of valence-shell polarizability.

Figure 6.27 illustrates the effect on the ^3P resonance of the net correlation energy parameter Δ (Thomas and Nesbet, 1975c). Values of Δ were chosen to give ^3P stabilized energy eigenvalues in the range 0–0.20 eV with respect to the computed target ground-state energy. The value $\Delta = 0.385$ eV places the resonance almost exactly on threshold. For the three values of Δ shown, the computed resonance parameters are

$$\Delta = 0.385 \text{ eV}, \quad E_{\text{res}} = 0.0 \text{ eV}, \qquad \Gamma = 0.0 \text{ eV},$$

$$\Delta = 0.575 \text{ eV}, \quad E_{\text{res}} = 0.1047 \text{ eV}, \quad \Gamma = 0.0209 \text{ eV}, \qquad (6.18)$$

$$\Delta = 0.950 \text{ eV}, \quad E_{\text{res}} = 0.3181 \text{ eV}, \quad \Gamma = 0.1309 \text{ eV}.$$

The single-configuration (SC) approximation gives $E_{\text{res}} = 0.89$ eV, $\Gamma = 0.64$ eV.

The computed elastic scattering cross sections are shown in Fig. 6.28 for several values of Δ (Thomas and Nesbet, 1975c) as solid curves. The dashed

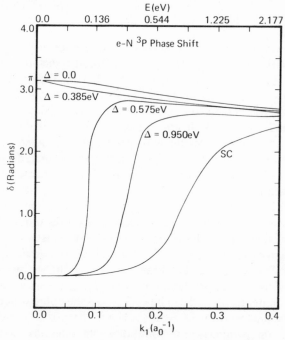

Figure 6.27. e^-–N, phase shift for ^3P scattering state (Thomas and Nesbet, 1975c, Fig. 1).

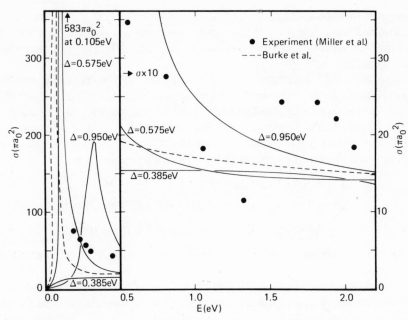

Figure 6.28. e^-–N, total elastic cross section (Thomas and Nesbet, 1975c, Fig. 2).

curve is the six-state R-matrix calculation of Burke *et al.* (1974), which places the ^3P resonance at 0.06 eV. The figure also includes experimental data points (heavy dots) of Miller *et al.* (1970). These data were not published in detail, and no error estimates were given. The value $\Delta = 0.575$ eV gives the best fit to the data of Miller *et al.* This would place the resonance peak at 0.105 eV, in good agreement with the value 0.12 eV from bound-state calculations (Moser and Nesbet, 1971). A somewhat larger value, 0.19 eV for the energy of $N^-(^3P)$ has been estimated by Sasaki and Yoshimine (1974), who extrapolated the residual error in elaborate variational calculations.

Figure 6.29 and 6.30, respectively, show the differential cross section computed at two energies for three values of Δ (Thomas and Nesbet, 1975a). As an aid to experimental determination of the true resonance energy, the best fit of differential cross-section data to these curves would determine the parameter Δ. Then E_{res} could be obtained from Eqs. (6.18) by interpolation.

The Rayleigh–Ritz variational principle is rigorously valid for the $^4S^0$ ground state of neutral N and can be expected to be approximately valid for a well-defined narrow resonance such as the ^3P state of N^- considered here. On this basis, the comparison of computed total energies shown in Figure 6.31 (Thomas and Nesbet, 1975c) indicates that the variational BG wave

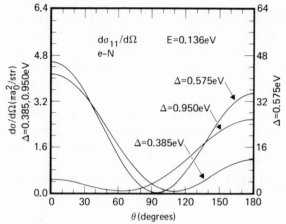

Figure 6.29. e^-–N, differential elastic cross section at 0.136 eV (Thomas and Nesbet, 1975c, Fig. 3).

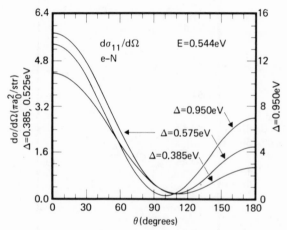

Figure 6.30. e^-–N, differential elastic cross section at 0.544 eV (Thomas and Nesbet, 1975c, Fig. 4).

functions are better solutions of Schrödinger's equation than any prior scattering calculations for all values of the parameter $\Delta = \Delta E$ considered. Energies obtained in prior electron scattering calculations are denoted in the figure by SC, CI, and Burke *et al.* (1974). Hartree–Fock energies of N and N^-, from pure bound-state calculations, are denoted by HF. The gap between the BG results and the best pure bound-state calculations, denoted by SY (Sasaki and Yoshimine, 1974), remains quite large. This is the essential reason for introducing the net correlation energy parameter Δ in the BG calculations.

Total and differential cross sections for processes connecting the $^4S^0$, $^2D^0$, and $^2P^0$ states were computed in the BG approximation, at level $[2s2p]_1$, with parameter $\Delta = 0.575$ eV for energies up to 9 eV (Thomas and Nesbet, 1975c). The excitation cross sections are shown in Fig. 6.32 as full lines compared with MCC results given as dashed lines (Ormonde *et al.*, 1973). The agreement is reasonable for $\sigma_{12}(^4S^0 \rightarrow {}^2D^0)$ and $\sigma_{13}(^4S^0 \rightarrow {}^2P^0)$. The MCC cross section $\sigma_{23}(^2D^0 \rightarrow {}^2P^0)$ apparently lacks the full effect of the $^3P^0$ scattering state, which dominates the rise from threshold and produces a shape or VS resonance near 10 eV. This resonance corresponds to the $N^-(^3P^0)$ state of configuration $2s2p^5$. It is also found near 10 eV in R-matrix calculations (Berrington *et al.*, 1975a; LeDourneuf *et al.*, 1975).

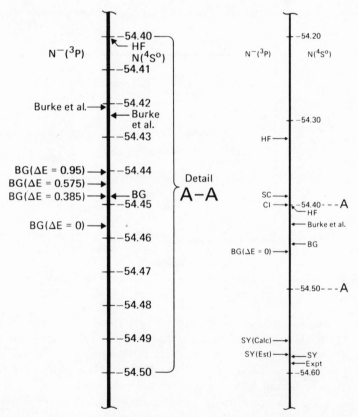

Figure 6.31. Comparison of ground-state energies of N and N$^-$ (Thomas and Nesbet, 1975c, Fig. 5).

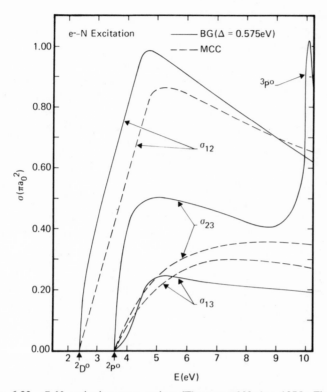

Figure 6.32. e^-–N, excitation cross sections (Thomas and Nesbet, 1975c, Fig. 7).

Figure 6.33 shows the full results of the BG calculations (with $\Delta = 0.575$ eV) in the form of a matrix of differential cross sections (Thomas and Nesbet, 1975c). Computed energy above the $^4S^0$ target atom ground state is given by $k^2/2$ in Hartree atomic units (27.212 eV).

There is no evidence in the scattering calculations of resonances associated with the expected states ^1D and ^1S of the N^- ground configuration $2s^2 2p^4$. This is compatible with the symmetry selection rules affecting these states if ^1D lies below the $^2D^0$ threshold (at 2.38 eV) and if ^1S lies below the $^2P^0$ threshold (at 3.58 eV). Extrapolation of excitation energies along the isoelectronic sequence Ne^{++}, F^+, O, N^- indicates that this would be the case if $N^-(^3P)$ occurs at 0.10 eV above $N(^4S^0)$. Hence the ^1D and ^1S states of N^- are expected to be metastable, decaying only through spin–spin or spin–orbit interactions (Thomas and Nesbet, 1975a). Tentative experimental evidence exists for the occurrence of long-lived N^- species in electron-impact excitation of NO and of N_2 (Hiraoka *et al.*, 1977).

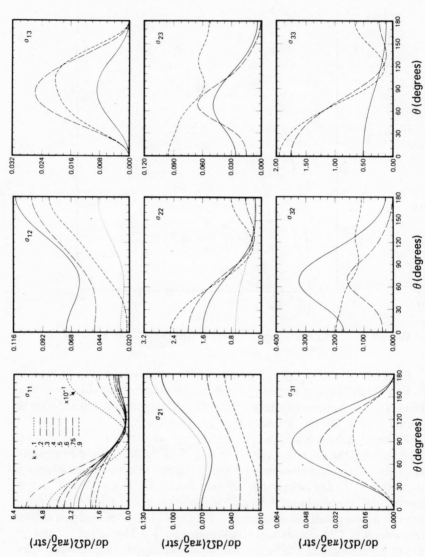

Figure 6.33. e^-–N, matrix of differential elastic and inelastic cross sections (Thomas and Nesbet, 1975c, Fig. 9).

Carbon

Like oxygen, carbon has a ^3P ground state and valence excited states ^1D and ^1S, but it also has a low-lying state $(2s2p^3)^5S^0$. The low-energy states of C^- are more complex than O^-, which has a single bound state $(2s^22p^5)^2P^0$ at -1.46 eV, relative to $O(^3P)$. The $2s^22p^3$ ground configuration of C^- has three states $^4S^0, ^2D^0$, and $^2P^0$. The lowest of these is bound at -1.27 eV, and the $^2D^0$ state is weakly bound (Ilin, 1973).

The remaining $^2P^0$ state of the C^- ground configuration would be expected to appear as a resonance in the ground-state scattering continuum of carbon. This C^- resonance should produce prominent structure in the scattering cross section, while the analogous O^- bound state is too far below threshold to have a strong influence.

Figure 6.34 shows variational SC and CI calculations (Thomas *et al.*, 1974b), compared with polarized orbital calculations of Henry (1968). The polarized orbital curve shows no resonance structure. The SC calculation shows resonance structure corresponding to a $^2D^0$ state of C^- near 0.5 eV and a $^2P^0$ state near 2 eV. The computed ^1S threshold in this approximation is indicated by (^1S) in the figure. The SC and polarized orbital cross sections are strikingly different. Both are qualitatively incorrect since the $^2D^0$ resonance should be a bound state just below threshold and the $^2P^0$ state should appear as a scattering resonance. The CI calculation is qualitatively correct, giving a $^2P^0$ resonance near 0.54 eV. The $^2D^0$ state has moved below threshold.

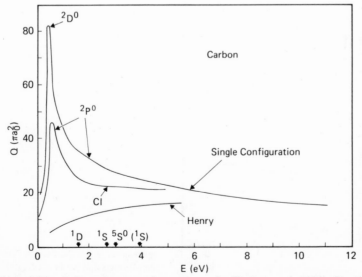

Figure 6.34. e^-–C, total cross section, early results (Thomas *et al.*, 1974b, Fig. 1).

Matrix variational calculations in the Bethe–Goldstone approximation (BG), at level $[2s2p]_1$, gave the total elastic cross section shown in Fig. 6.35 (Thomas and Nesbet, 1975d). The polarized orbital cross section (Henry, 1968) is shown for comparison. A net correlation correction parameter was included in the BG calculations. Its value, $\Delta = 0.530$ eV, was determined by adjusting the $^2D^0$ energy to its experimental value. The same value of Δ was used for both $^2D^0$ and $^2P^0$ components of the scattering wave function. With this value of Δ, the $^2P^0$ resonance peak, shown in Fig. 6.35, is at 0.461 eV, and the resonance width is 0.233 eV. In the absence of experimental data, these results can be taken to predict a $^2P^0$ resonance peak in e^-–C scattering between 0.4 and 0.6 eV.

Excitation cross sections computed in the BG approximation, with $\Delta = 0.530$ eV, shown as full curves in Fig. 6.36, are compared with close-coupling results in the SC approximation (Henry *et al.*, 1969), shown as dashed curves. There is reasonable agreement for $\sigma_{13}(^3P \to {}^1S)$ and $\sigma_{23}(^1D \to {}^1S)$. The leading peak in $\sigma_{12}(^3P \to {}^1D)$ in the SC calculations is twice as high as the BG peak. This is probably due to the $^2P^0$ resonance, which is displaced upward to about 2 eV in the SC approximation. Differential cross sections for all processes up to total energy 7 eV were computed by the BG method (Thomas and Nesbet, 1975d). Similar results have been obtained in the polarized frozen core approximation, using the R-matrix method (LeDourneuf, 1976).

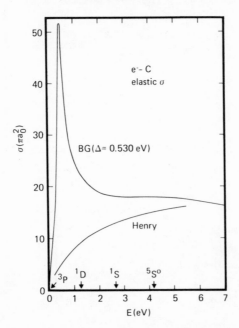

Figure 6.35. e^-–C total elastic cross section (Thomas and Nesbet, 1975d, Fig. 1).

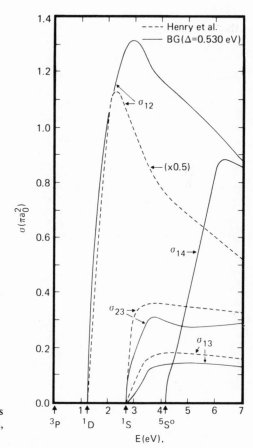

Figure 6.36. $e^- –C$ excitation cross sections (Thomas and Nesbet, 1975d, Fig. 3).

References

Abdallah, J., Jr., and Truhlar, D. G. (1974), *J. Chem. Phys.* **60,** 4670.

Abdallah, J., Jr., and Truhlar, D. G. (1975), *Comput. Phys. Commun.* **9,** 327.

Abdel-Raouf, M. A. (1979), *J. Phys. B* **12,** 3349.

Abdel-Raouf, M. A., and Belschner, D. (1978), *J. Phys. B* **11,** 3677.

Abouaf, R., and Teillet-Billy, D. (1977), *J. Phys. B* **10,** 2261.

Abramowitz, M., and Stegun, I. A. (1964), Eds. *Handbook of Mathematical Functions, National Bureau of Standards Applied Mathematics Series No. 55,* U.S. Government Printing Office, Washington, D.C.

Ahlrichs, R., Lischka, H., Staemmler, V., and Kutzelnigg, W. (1975), *J. Chem. Phys.* **62,** 1225.

Andrick, D. (1973), *Adv. At. Mol. Phys.* **9,** 207.

Andrick, D. (1979), *J. Phys. B* **12,** L175.

Andrick, D., and Bitsch, A. (1975), *J. Phys. B* **8,** 393.

Armstead, R. L. (1968), *Phys. Rev.* **171,** 91.

Bardsley, J. N., and Nesbet, R. K. (1973), *Phys. Rev. A* **8,** 203.

Bederson, B. (1970a), *Comments At. Mol. Phys.* **1,** 135.

Bederson, B. (1970b), *Comments At. Mol. Phys.* **2,** 7.

Bederson, B., and Kieffer, L. J. (1971), *Rev. Mod. Phys.* **43,** 601.

Berrington, K. A., Burke, P. G., and Robb, W. D. (1975a), *J. Phys. B* **8,** 2500.

Berrington, K. A., Burke, P. G., and Sinfailam, A. L. (1975b), *J. Phys. B* **8,** 1459.

Bethe, H. A., and Goldstone, J. (1957), *Proc. Roy. Soc (London) A* **238,** 551.

Birtwistle, D. T., and Herzenberg, A. (1971), *J. Phys. B* **4,** 53.

Blaha, M., and Davis, J. (1975), *Phys. Rev. A* **12,** 2319.

Blatt, J. M., and Biedenharn, L. C. (1952), *Rev. Mod. Phys.* **24,** 258.

Bloch, C. (1957), *Nucl. Phys.* **4,** 503.

Bottcher, C. (1970), *J. Comput. Phys.* **6,** 237.

Brandsden, B. H. (1970), *Atomic Collision Theory,* Benjamin, New York.

Breit, G. (1959), *Handbuch der Physik,* Vol. 41, No. 1, Springer, Berlin.

Brenig, W., and Haag, R. (1959), *Fortschr. Physik* **7,** 183.

Brink, D. M., and Satchler, G. R. (1968), *Angular Momentum,* 2nd ed., Oxford University Press, London.

Brode, R. (1929), *Phys. Rev.* **34,** 673.

Brongersma, H. H., Knoop, F. W. E., and Backx, C. (1972), *Chem. Phys. Lett.* **13,** 16.

Brownstein, K. R., and McKinley, W. A. (1968), *Phys. Rev.* **170,** 1255.

Brueckner, K. A. (1959), in *The Many-Body Problem* (B. de Witt, ed.), Wiley, New York, p. 47.

Brunt, J. N. H., King, G. C., and Read, F. H. (1977), *J. Phys. B* **10**, 433.

Burke, P. G. (1965), *Adv. Phys.* **14**, 521.

Burke, P. G. (1968), *Adv. At. Mol. Phys.* **4**, 173.

Burke, P. G. (1973), *Comput. Phys. Commun.* **6**, 288.

Burke, P. G. (1977), *Potential Scattering in Atomic Physics*, Plenum Press, New York.

Burke, P. G., and Robb, W. D. (1972), *J. Phys. B* **5**, 44.

Burke, P. G., and Robb, W. D. (1975), *Adv. At. Mol. Phys.* **11**, 143.

Burke, P. G., and Seaton, M. J. (1971), *Methods Comput. Phys.* **10**, 1.

Burke, P. G., and Schey, H. M. (1962), *Phys. Rev.* **126**, 147.

Burke, P. G., and Taylor, A. J. (1969), *J. Phys. B* **2**, 869.

Burke, P. G., Ormonde, S., and Whitaker, W. (1967), *Proc. Phys. Soc. (London)* **92**, 319.

Burke, P. G., Cooper, J. W., and Ormonde, S. (1969a), *Phys. Rev.* **183**, 245.

Burke, P. G., Gallaher, D. F., and Geltman, S. (1969b), *J. Phys. B* **2**, 1142.

Burke, P. G., Hibbert, A., and Robb, W. D. (1971), *J. Phys. B* **4**, 1153.

Burke, P. G., Berrington, K. A., LeDourneuf, M., and Vo Ky Lan (1974), *J. Phys. B* **7**, L531.

Buttle, P. J. A. (1967), *Phys. Rev.* **160**, 719.

Callaway, J. (1973), *Comput. Phys. Commun.* **6**, 265.

Callaway, J. (1978a), *Phys. Lett.* **65A**, 199.

Callaway, J. (1978b), *Phys. Reports* **45**, 89.

Callaway, J., and Williams, J. F. (1975), *Phys. Rev. A* **12**, 2312.

Callaway, J., and Wooten, J. W. (1974), *Phys. Rev. A* **9**, 1924.

Callaway, J., and Wooten, J. W. (1975), *Phys. Rev. A* **11**, 1118.

Callaway, J., La Bahn, R. W., Pu, R. T., and Duxler, W. M. (1968), *Phys. Rev.* **168**, 12.

Callaway, J., McDowell, M. R. C., and Morgan, L. A. (1975), *J. Phys. B* **8**, 2181.

Callaway, J., McDowell, M. R. C., and Morgan, L. A. (1976), *J. Phys. B* **9**, 2043.

Castillejo, L., Percival, I. C., and Seaton, M. J. (1960), *Proc. Roy. Soc. (London) A* **254**, 259.

Chandra, N., and Temkin, A. (1976), *Phys. Rev. A* **13**, 188.

Chang, E. S., and Temkin, A. (1970), *J. Phys. Soc. Japan* **29**, 172.

Chase, D. M. (1956), *Phys. Rev.* **104**, 838.

Chatwin, R. A., and Purcell, J. E. (1971), *J. Math. Phys.* **12**, 2024.

Chen, J. C. Y. (1967), *Phys. Rev.* **156**, 150.

Chen, J. C. Y. (1970), *Nucl. Instrum. Methods* **90**, 237.

Chen, J. C. Y., and Chung, K. T. (1970), *Phys. Rev. A* **2**, 1892.

Chen, J. C. Y., Chung, K. T., and Sinfailam, A. L. (1971), *Phys. Rev. A* **4**, 1517.

Chung, K. T., and Hurst, R. P. (1966), *Phys. Rev.* **152**, 35.

Cizek, J., and Paldus, J. (1971), *Int. J. Quantum Chem.* **5**, 359.

Collins, R. E., Bederson, B., and Goldstein, M. (1971), *Phys. Rev. A* **3**, 1976.

Condon, E. U., and Shortley, G. H. (1935), *The Theory of Atomic Spectra*, Cambridge University Press, New York.

Crompton, R. W., Elford, M. T., and Jory, R. L. (1967), *Austral. J. Phys.* **20**, 369.

Crompton, R. W., Elford, M. T., and Robertson, A. G. (1970), *Austral. J. Phys.* **23**, 667.

Csanak, Gy., and Taylor, H. S. (1972), *Phys. Rev. A* **6**, 1843.

Csanak, Gy., and Taylor, H. S. (1973), *J. Phys. B* **6**, 2055.

Csanak, Gy., Taylor, H. S., and Yaris, R. (1971a), *Phys. Rev. A* **3**, 1322.

Csanak, Gy., Taylor, H. S., and Yaris, R. (1971b), *Adv. At. Mol. Phys.* **7**, 287.

Csanak, Gy., Taylor, H. S., and Tripathy, D. N. (1973), *J. Phys. B* **6**, 2040.

Cvejanovic, S., and Read, F. H. (1974), *J. Phys. B* **7**, 1180.

Cvejanovic, S., Comer, J., and Read, F. H. (1974), *J. Phys. B* **7**, 468.

Dalitz, R. H. (1961), *Rev. Mod. Phys.* **33**, 471.

Damburg, R. J., and Geltman, S. (1968), *Phys. Rev. Lett.* **20**, 485.

Damburg, R. J., and Karule, E. (1967), *Proc. Phys. Soc. (London)* **90**, 637.

Das, J. N., and Rudge, M. R. H. (1976), *J. Phys. B* **9**, L131.

Dehmel, R. C., Fineman, M. A., and Miller, D. R. (1974), *Phys. Rev. A* **9**, 1564.

Dehmel, R. C., Fineman, M. A., and Miller, D. R. (1976), *Phys. Rev. A* **13**, 115.

Demkov, Yu. N. (1963), *Variational Principles in the Theory of Collisions* (N. Kemmer, trans.), Pergamon Press, London.

Demkov, Yu. N., and Shepelenko, F. P. (1958), *Sov. Phys. JETP* **6**, 1144.

Drachman, R. J., and Temkin, A. (1972), in *Case Studies in Atomic Collision Physics*, Vol. 2 (E. W. McDaniel and M. R. C. McDowell, eds.), North-Holland, Amsterdam, p. 399.

Drake, G. W. F. (1970). *Phys. Rev. Lett.* **24**, 126.

Duxler, W. M., Poe, R. T., and La Bahn, R. W. (1971), *Phys. Rev. A* **4**, 1935.

Eckart, C. (1930), *Rev. Mod. Phys.* **2**, 305.

Edmonds, A. R. (1957), *Angular Momentum in Quantum Mechanics*, Princeton University Press, Princeton, New Jersey.

Ehrhardt, H., and Willmann, K. (1967a), *Z. Phys.* **203**, 1.

Ehrhardt, H., and Willmann, K. (1967b), *Z. Phys.* **204**, 462.

Ehrhardt, H., Langhans, L., and Linder, F. (1968), *Z. Phys.* **214**, 179.

Eissner, W., and Seaton, M. J. (1972), *J. Phys. B* **5**, 2187.

Fano, U. (1961), *Phys. Rev.* **124**, 1866.

Fano, U. (1965), *Phys. Rev.* **140**, A67.

Fano, U., and Cooper, J. W. (1965), *Phys. Rev.* **137**, 1364.

Fano, U., and Lee, C. M. (1973), *Phys. Rev. Lett.* **31**, 1573.

Fels, M. F., and Hazi, A. U. (1970), *Phys. Rev. A* **1**, 1109.

Fels, M. F., and Hazi, A. U. (1971), *Phys. Rev. A* **4**, 662.

Fels, M. F., and Hazi, A. U. (1972), *Phys. Rev. A* **5**, 1236.

Feshbach, H. (1958), *Ann. Phys. (New York)* **5**, 357.

Feshbach, H. (1962), *Ann. Phys. (New York)* **19**, 287.

Fliflet, A. W., and McKoy, V. (1978), *Phys. Rev. A* **18**, 2107.

Fon, W. C., Berrington, K. A., Burke, P. G., and Kingston, A. E. (1978), *J. Phys. B* **11**, 325.

Frazer, R. A., Duncan, W. J., and Collar, A. R. (1947), *Elementary Matrices*, Cambridge University Press, New York.

Gailitis, M. (1965a), *Sov. Phys JETP* **20**, 107. [(1964), *Zh. Eksp. Teor. Fiz.* **47**, 160.]

Gailitis, M. (1965b), in *Abstracts, Fourth International Conference on the Physics of Electronic and Atomic Collisions*, Science Bookcrafters, Hastings-on-Hudson, New York, p. 10.

Gailitis, M., and Damburg, R. (1963a), *Sov. Phys. JETP* **17**, 1107.

Gailitis, M., and Damburg, R. (1963b), *Proc. Phys. Soc. (London)* **82**, 192.

Geltman, S. (1969), *Topics in Atomic Collision Theory*, Academic Press, New York.

Geltman, S., and Burke, P. G. (1970), *J. Phys. B* **3**, 1062.

Gibson, R. J., and Dolder, K. T. (1969), *J. Phys. B* **2**, 741.

Golden, D. E., and Bandel, H. W. (1965), *Phys. Rev.* **138**, A14.

Golden, D. E., and Zecca, A. (1971), *Rev. Sci. Instrum.* **42**, 210.

Golden, D. E., Lane, N. F., Temkin, A., and Gerjuoy, E. (1971), *Rev. Mod. Phys.* **43**, 642.

Gomes, L. C., Walecka, J. D., and Weisskopf, V. F. (1958), *Ann. Phys. (New York)* **3**, 241.

Hahn, Y. (1971), *Phys. Rev. A* **4**, 1881.

Hahn, Y., and Spruch, L. (1967), *Phys. Rev.* **153**, 1159.

Hahn, Y., O'Malley, T. F., and Spruch, L. (1962), *Phys. Rev.* **128**, 932.

Hahn, Y., O'Malley, T. F., and Spruch, L. (1963), *Phys. Rev.* **130**, 381.

Hahn, Y., O'Malley, T. F., and Spruch, L. (1964a), *Phys. Rev.* **134**, B397.

Hahn, Y., O'Malley, T. F., and Spruch, L. (1964b), *Phys. Rev.* **134**, B911.
Hall, R. I., Reinhardt, J., Joyez, G., and Mazeau, J. (1972), *J. Phys. B* **5**, 66.
Harris, F. E. (1967), *Phys. Rev. Lett.* **19**, 173.
Harris, F. E., and Michels, H. H. (1969a), *Phys. Rev. Lett.* **22**, 1036.
Harris, F. E., and Michels, H. H. (1969b), *J. Comput. Phys.* **6**, 237.
Harris, F. E., and Michels, H. H. (1971), *Methods Comput. Phys.* **10**, 143.
Hartree, D. R. (1957), *The Calculation of Atomic Structures*, Wiley, New York.
Hazi, A. U. (1978), *J. Phys. B* **11**, L259.
Hazi, A. U., and Fels, M. F. (1971), *Chem. Phys. Lett.* **8**, 582.
Hazi, A. U., and Taylor, H. S. (1970), *Phys. Rev. A* **1**, 1109.
Henry, R. J. W. (1967), *Phys. Rev.* **162**, 56.
Henry, R. J. W. (1968), *Phys. Rev.* **172**, 99.
Henry, R. J. W., Burke, P. G., and Sinfailam, A. L. (1969), *Phys. Rev.* **178**, 218.
Hibbert, A. (1975), *Comput. Phys. Commun.* **9**, 141.
Hiraoka, H., Nesbet, R. K., and Welsh, L. W., Jr. (1977), *Phys. Rev. Lett.* **39**, 130.
Hoegy, W. R. (1976), *Geophys. Res. Lett.* **3**, 541.
Holøien, E. (1960), *J. Chem. Phys.* **33**, 301.
Householder, A. S. (1964), *The Theory of Matrices in Numerical Analysis*, Blaisdell, New York.
Hulthén, L. (1944), *Kgl. Fysiogr. Sällsk. Lund Förh.* **14**, No. 21.
Hulthén, L. (1948), *Arkiv Mat. Astron. Fysik* **35A**, No. 25.
Hylleraas, E. A. (1928), *Z. Phys.* **48**, 469.
Hylleraas, E. A., and Undheim, B. (1930), *Z. Phys.* **65**, 759.
Ilin, R. N. (1973), in *Atomic Physics 3* (S. J. Smith and G. K. Walters, eds.), Plenum Press, New York, p. 309.
Jackson, J. L. (1951), *Phys Rev.* **83**, 301.
Jacob, M., and Wick, G. C. (1959), *Ann. Phys. (New York)* **7**, 404.
Joachain, C. J. (1975), *Quantum Collision Theory*, North-Holland, Amsterdam.
Karule, E. (1965), in *Cross Sections of Electron–Atom Collisions* (V. Veldre, ed.), Latvian Academy of Sciences, Riga, p. 33.
Karule, E. (1972), *J. Phys. B* **5**, 2051.
Karule, E., and Peterkop, R. K. (1965), in *Cross Sections of Electron–Atom Collisions* (V. Veldre, ed.), Latvian Academy of Sciences, Riga, p. 3.
Kasdan, A., Miller, T. M., and Bederson, B. (1973), *Phys. Rev. A* **8**, 1562.
Kato, T. (1950), *Phys. Rev.* **80**, 475.
Kato, T. (1951), *Prog. Theor. Phys.* **6**, 394.
Kauppila, W. E., Stein, T. S., Jesion, G., Dababneh, M. S., and Pol, V. (1977), *Rev. Sci. Instrum.* **48**, 322.
Kelly, H. P. (1967), *Phys. Rev.* **160**, 44.
Kelly, H. P. (1968), *Phys. Rev.* **171**, 54.
Kennerly, R. E., and Bonham, R. A. (1978), *Phys. Rev. A* **17**, 1844.
Knowles, M., and McDowell, M. R. C. (1973), *J. Phys. B* **6**, 300.
Kohn, W. (1948), *Phys. Rev.* **74**, 1763.
Koopmans, T. (1933), *Physica* **1**, 104.
Kutzelnigg, W. (1977), in *Methods of Electronic Structure Theory* (H. F. Schaefer III, ed.), Plenum Press, New York, p. 129.
Lane, A. M., and Robson, D. (1969), *Phys. Rev.* **178**, 1715.
Lane, A. M., and Thomas, R. G. (1958), *Rev. Mod. Phys.* **30**, 257.
Langhoff, P. W. (1974), *Int. J. Quant. Chem. Symp.* **8**, 347.

Langhoff, P. W., and Corcoran, C. T. (1974), *J. Chem. Phys.* **61**, 146.

LeDourneuf, M. (1976), Thesis, University of Paris VI (CNRS Rept. A012658).

LeDourneuf, M., and Nesbet, R. K. (1976), *J. Phys. B* **9**, L241.

LeDourneuf, M., Vo Ky Lan, Berrington, K. A., and Burke, P. G. (1975), in *Abstracts, Ninth International Conference on the Physics of Electronic and Atomic Collisions*, University of Washington Press, Seattle, p. 634.

LeDourneuf, M., van Regemorter, H., and Vo Ky Lan (1976), in *Electron and Photon Interactions with Atoms* (H. Kleinpoppen and M. R. C. McDowell, eds.), Plenum Press, New York, p. 415.

LeDourneuf, M., Vo Ky Lan, and Burke, P. G. (1977), *Comments At. Mol. Phys.* **7**, 1.

Lee, C. M. (1974), *Phys. Rev. A* **10**, 584.

Leonard, P. J., and Barker, J. A. (1975), in *Theoretical Chemistry, Advances and Perspectives*, Vol. 1 (H. Eyring and D. Henderson, eds.), Academic Press, New York, p. 117.

Levy, B. R., and Keller, J. B. (1963), *J. Math. Phys.* **4**, 54.

Lin, S. C., and Kivel, B. (1959), *Phys. Rev.* **114**, 1026.

Lippmann, B. A., and Schwinger, J. (1950), *Phys. Rev.* **79**, 469.

Löwdin, P. -O. (1955), *Phys. Rev.* **97**, 1509.

Lyons, J. D., and Nesbet, R. K. (1969), *J. Comput. Phys.* **4**, 499.

Lyons, J. D., and Nesbet, R. K. (1973), *J. Comput. Phys.* **11**, 166.

Lyons, J. D., Nesbet, R. K., Rankin, C. C., and Yates, A. C. (1973), *J. Comput. Phys.* **13**, 229.

Macek, J., (1970), *Phys. Rev. A* **2**, 1101.

Macek, J., and Burke, P. G. (1967), *Proc. Phys. Soc. (London)* **92**, 351.

Malik, F. B. (1962), *Ann. Phys. (New York)* **20**, 464.

Martin, W. C. (1973), *J. Phys. Chem. Ref. Data* **2**, 257.

Matese, J. J., and Oberoi, R. S. (1971), *Phys. Rev. A* **4**, 569.

McDowell, M. R. C., Myerscough, V. P., and Narain, U. (1974), *J. Phys. B* **7**, L195.

McGowan, J. W. (1967), *Phys. Rev.* **156**, 165.

McGowan, J. W., Williams, J. F., and Curley, E. K. (1969), *Phys. Rev.* **180**, 132.

McVoy, K. (1967), in *Fundamentals of Nuclear Theory* (A. de-Shalit and C. Villi, eds.), IAEA, Vienna, p. 419.

Meyer, W. (1974), *Theor. Chim. Acta* **35**, 277.

Michels, H. H., Harris, F. E., and Scolsky, R. N. (1969), *Phys. Lett.* **28A**, 467.

Miller, W. H. (1970), *Chem. Phys. Lett.* **4**, 627.

Miller, T. M., and Bederson, B. (1977), *Adv. At. Mol. Phys.* **13**, 1.

Miller, T. M., Aubrey, B. B., Eisner, P. N., and Bederson, B. (1970), *Bull. Am. Phys. Soc.* **15**, 416.

Milloy, H. B., and Crompton, R. W. (1977), *Phys. Rev. A* **15**, 1847.

Mittleman, M. H. (1966), *Phys. Rev.* **147**, 69.

Moiseiwitsch, B. L. (1966), *Variational Principles*, Wiley-Interscience, New York.

Moores, D. L. and Norcross, D. W. (1972), *J. Phys. B* **5**, 1482.

Morawitz, H. (1970), *J. Math. Phys.* **11**, 649.

Morgan, L. A., McDowell, M. R. C., and Callaway, J. (1977), *J. Phys. B* **10**, 3297.

Moser, C. M., and Nesbet, R. K. (1971), *Phys. Rev. A* **4**, 1336.

Mott, N. F., and Massey, H. S. W. (1965), *The Theory of Atomic Collisions*, Oxford University Press, New York.

Nesbet, R. K. (1961), *J. Math. Phys.* **2**, 701.

Nesbet, R. K. (1963), *Rev. Mod. Phys.* **35**, 552.

Nesbet, R. K. (1965), *Adv. Chem. Phys.* **9**, 321.

Nesbet, R. K. (1967), *Phys. Rev.* **156,** 99.

Nesbet, R. K. (1968), *Phys. Rev.* **175,** 134.

Nesbet, R. K. (1969a), *Phys. Rev.* **179,** 60.

Nesbet, R. K. (1969b), *Adv. Chem. Phys.* **14,** 1.

Nesbet, R. K. (1970), *Phys. Rev. A* **2,** 661.

Nesbet, R. K. (1971), *J. Comput. Phys.* **8,** 483.

Nesbet, R. K. (1973), *Bull. Am. Phys. Soc.* **18,** 1501.

Nesbet, R. K. (1975a), *Adv. Quantum Chem.* **9,** 215.

Nesbet, R. K. (1975b), *Phys. Rev. A* **12,** 444.

Nesbet, R. K. (1977a), *Adv. At. Mol. Phys.* **13,** 315.

Nesbet, R. K. (1977b), *J. Phys. B* **10,** L739.

Nesbet, R. K. (1977c), *Comments At. Mol. Phys.* **7,** 15.

Nesbet, R. K. (1978a), *J. Phys. B* **11,** L21.

Nesbet, R. K. (1978b), *Phys. Rev. A* **18,** 955.

Nesbet, R. K. (1978c), in *Proceedings, 1978 IBM Symposium on Mathematics and Computation* (K. L. Deckert, ed.), IBM, San Jose, California, p. 11.

Nesbet, R. K. (1979a), *Phys. Rev. A* **19,** 551.

Nesbet, R. K. (1979b), *J. Phys. B* **12,** L243.

Nesbet, R. K. (1979c), *Phys. Rev. A* **20,** 58.

Nesbet, R. K., and Lyons, J. D. (1971), *Phys. Rev. A* **4,** 1812.

Nesbet, R. K., and Oberoi, R. S. (1972), *Phys. Rev. A* **6,** 1855.

Nesbet, R. K., Oberoi, R. S., and Bardsley, J. N. (1974), *Chem. Phys. Lett.* **25,** 587.

Newton, R. G. (1966), *Scattering Theory of Waves and Particles,* McGraw-Hill, New York.

Neynaber, R. H., Marino, L. L., Rothe, E. W., and Trujillo, S. M. (1961), *Phys. Rev.* **123,** 148.

Neynaber, R. H., Marino, L. L., Rothe, E. W., and Trujillo, S. M. (1963), *Phys. Rev.* **129,** 2069.

Norcross, D. W. (1969), *J. Phys. B* **2,** 1300.

Norcross, D. W. (1971), *J. Phys. B* **4,** 1458.

Norcross, D. W., and Moores, D. L. (1973), in *Atomic Physics* (S. J. Smith and G. K. Walters, eds.), Plenum Press, New York, p. 261.

Nordholm, S., and Bacskay, G. (1978), *J. Phys. B* **11,** 193.

Nuttall, J. (1969), *Ann. Phys. (New York)* **52,** 428.

Oberoi, R. S., and Nesbet, R. K. (1973a), *Phys. Rev. A* **8,** 215.

Oberoi, R. S., and Nesbet, R. K. (1973b), *J. Comput. Phys.* **12,** 526.

Oberoi, R. S., and Nesbet, R. K. (1973c), *Phys. Rev. A* **8,** 2969.

Oberoi, R. S., and Nesbet, R. K. (1974), *Phys. Rev. A* **9,** 2804.

Oberoi, R. S., Callaway, J., and Seiler, G. J. (1972), *J. Comput. Phys.* **10,** 466.

O'Malley, T. F. (1963), *Phys. Rev.* **130,** 1020.

O'Malley, T. F., Spruch, L., and Rosenberg, L. (1961), *J. Math. Phys.* **2,** 491.

O'Malley, T. F., Burke, P. G., and Berrington, K. A. (1979), *J. Phys. B* **12,** 953.

Ormonde, S., Smith, K., Torres, B. W., and Davis, A. R. (1973), *Phys. Rev. A* **8,** 262.

Patterson, T. A., Hotop, H., Kasdan, A., Norcross, D. W., and Lineberger, W. C. (1974), *Phys. Rev. Lett.* **32,** 189.

Penrose, R. (1955), *Proc. Camb. Phil. Soc.* **51,** 406.

Percival, I. C., and Seaton, M. J. (1957), *Proc. Camb. Phil. Soc.* **53,** 654.

Perel, J., Englander, P., and Bederson, B. (1962), *Phys. Rev.* **128,** 1148.

Pichou, F., Huetz, A., Joyez, G., Landau, M., and Mazeau, J. (1976), *J. Phys. B* **9,** 933.

Pu, R. T., and Chang, E. (1966), *Phys. Rev.* **151,** 31.

Purcell, J. E. (1969), *Phys. Rev.* **185,** 1279.

Ramaker, D. E. (1972), *J. Math. Phys.* **13,** 161.

Register, D., and Poe, R. T. (1975), *Phys. Lett.* **51A,** 431.

Rescigno, T. N., McCurdy, C. W., and McKoy, V. (1974a), *Phys. Rev. Lett.* **27**, 401.

Rescigno, T. N., McCurdy, C. W., and McKoy, V. (1974b), *Phys. Rev. A* **10**, 2240.

Rescigno, T. N., McCurdy, C. W., and McKoy, V. (1975), *Phys. Rev. A* **11**, 825.

Risley, J. S., Edwards, A. K., and Geballe, R. (1974), *Phys. Rev. A* **9**, 1115.

Rohr, K., and Linder, F. (1975), *J. Phys. B* **8**, L200.

Rohr, K., and Linder, F. (1976), *J. Phys. B* **9**, 2521.

Rose, M. E. (1957), *Elementary Theory of Angular Momentum*, Wiley, New York.

Rountree, S. P., and Parnell, G. (1977), *Phys. Rev. Lett.* **39**, 853.

Rountree, S. P., Smith, E. R., and Henry, R. J. W. (1974), *J. Phys. B* **7**, L167.

Rubinow, S. I. (1955), *Phys. Rev.* **98**, 183.

Rudge, M. R. H. (1973), *J. Phys. B* **6**, 1788.

Rudge, M. R. H. (1975), *J. Phys. B* **8**, 940.

Sanche, L., and Burrow, P. D. (1972), *Phys. Rev. Lett.* **29**, 1639.

Sanche, L., and Schulz, G. J. (1972), *Phys. Rev. A* **4**, 1672.

Saraph, H. E. (1972), *Comput. Phys. Commun.* **3**, 256.

Saraph, H. E. (1973), *J. Phys. B* **6**, L243.

Sasaki, F., and Yoshimine, M. (1974), *Phys. Rev. A* **9**, 26.

Schlessinger, L., and Payne, G. L. (1974), *Phys. Rev. A* **10**, 1559.

Schneider, B., Taylor, H. S., and Yaris, R. (1970), *Phys. Rev. A* **1**, 855.

Schulz, G. J. (1973a), *Rev. Mod. Phys.* **45**, 378.

Schulz, G. J. (1973b), *Rev. Mod. Phys.* **45**, 423.

Schwartz, C. (1961a), *Ann. Phys. (New York)* **16**, 36.

Schwartz, C. (1961b), *Phys. Rev.* **124**, 1468.

Schwinger, J. (1947), *Phys. Rev.* **72**, 742; unpublished notes, Harvard University.

Schwinger, J. (1951), *Proc. Natl. Acad. Sci. USA* **37**, 452.

Seaton, M. J. (1953), *Philos. Trans. Roy. Soc. (London) A* **245**, 469.

Seaton, M. J. (1966), *Proc. Phys. Soc. (London)* **89**, 469.

Seaton, M. J. (1973), *Comput. Phys. Commun.* **6**, 247.

Seaton, M. J. (1974), *J. Phys. B* **7**, 1817.

Seiler, G. J., Oberoi, R. S., and Callaway, J. (1971), *Phys. Rev. A* **3**, 2006.

Shimamura, I. (1971a), *J. Phys. Soc. Japan* **30**, 1702.

Shimamura, I. (1971b), *J. Phys. Soc. Japan* **31**, 852.

Shimamura, I. (1977), in *The Physics of Electronic and Atomic Collisions* (Invited papers, X ICPEAC, Paris, 1977), North-Holland, Amsterdam.

Sinfailam, A. L. (1976), *J. Phys. B* **9**, L101.

Sinfailam, A. L., and Nesbet, R. K. (1972), *Phys. Rev. A* **6**, 2118.

Sinfailam, A. L., and Nesbet, R. K. (1973), *Phys. Rev. A* **7**, 1987.

Smith, K. (1971), *The Calculation of Atomic Collision Processes*, Wiley-Interscience, New York.

Smith, R. L., and Truhlar, D. G. (1973), *Comput. Phys. Commun.* **5**, 80.

Smith, K., Henry, R. J. W., and Burke, P. G. (1967), *Phys. Rev.* **157**, 51.

Stein, T. S., Kauppila, W. E., Pol, V., Smart, J. H., and Jesion, G. (1978), *Phys. Rev. A* **17**, 1600.

Sugar, R., and Blankenbecler, R. (1964), *Phys Rev.* **136**, B472.

Sunshine, G., Aubrey, B. B., and Bederson, B. (1967), *Phys. Rev.* **154**, 1.

Takatsuka, K., and Fueno, T. (1979a), *Phys. Rev. A* **19**, 1011.

Takatsuka, K., and Fueno, T. (1979b), *Phys. Rev. A* **19**, 1018.

Tambe, B. R., and Henry, R. J. W. (1974), *Phys. Rev. A* **10**, 2087.

Tambe, B. R., and Henry, R. J. W. (1976a), *Phys. Rev. A* **13**, 224.

Tambe, B. R., and Henry, R. J. W. (1976b), *Phys. Rev. A* **14**, 512.

Taylor, A. J., and Burke, P. G. (1967), *Proc. Phys. Soc. (London)* **92**, 336.
Taylor, H. S. (1970), *Adv. Chem. Phys.* **18**, 91.
Taylor, H. S., and Hazi, A. U. (1976), *Phys. Rev. A* **14**, 2071.
Taylor, H. S., and Thomas, L. D. (1972), *Phys. Rev. Lett.* **28**, 1091.
Temkin, A. (1957), *Phys. Rev.* **107**, 1004.
Temkin, A. (1962), *Phys. Rev.* **126**, 130.
Temkin, A., and Lamkin, J. C. (1961), *Phys. Rev.* **121**, 788.
Temkin, A., Bhatia, A. K., and Bardsley, J. N. (1972), *Phys. Rev. A* **5**, 1663.
Thomas, L. D., and Nesbet, R. K. (1974), in *Abstracts, Fourth International Conference on Atomic Physics*, Heidelberg, p. 468.
Thomas, L. D., and Nesbet, R. K. (1975a), *Phys. Rev. A* **11**, 170.
Thomas, L. D., and Nesbet, R. K. (1975b), *Phys. Rev. A* **12**, 1729.
Thomas, L. D., and Nesbet, R. K. (1975c), *Phys. Rev. A* **12**, 2369.
Thomas, L. D., and Nesbet, R. K. (1975d), *Phys. Rev. A* **12**, 2378.
Thomas, L. D., and Nesbet, R. K. (1975e), in *Abstracts, Ninth International Conference on the Physics of Electronic and Atomic Collisions*, University of Washington Press, Seattle, p. 637.
Thomas, L. D., Yarlagadda, B. S., Csanak, Gy., and Taylor, H. S. (1973), *Comput. Phys. Commun.* **6**, 316.
Thomas, L. D., Csanak, Gy., Taylor, H. S., and Yarlagadda, B. S. (1974a), *J. Phys. B* **7**, 1719.
Thomas, L. D., Oberoi, R. S., and Nesbet, R. K. (1974b), *Phys. Rev. A* **10**, 1605.
Thompson, D. G. (1966), *Proc. Roy. Soc. (London)* **A294**, 160.
Trajmar, S. (1973), *Phys. Rev. A* **8**, 191.
Truhlar, D. G., Abdallah, J., and Smith, R. L. (1974), *Adv. Chem. Phys.* **25**, 211.
Vo Ky Lan (1971), *J. Phys. B* **4**, 658.
Vo Ky Lan, Feutrier, N., Le Dourneuf, M., and van Regemorter, H. (1972), *J. Phys. B* **5**, 1506.
Vo Ky Lan, Le Dourneuf, M., and Burke, P. G. (1976), *J. Phys. B* **9**, 1065.
Walton, D. S., Peart, B., and Dolder, K. (1970), *J. Phys. B* **3**, L148.
Wichmann, E., and Heiss, P. (1974), *J. Phys. B* **7**, 1042.
Wigner, E. P. (1927), *Z. Phys.* **43**, 624.
Wigner, E. P. (1948), *Phys. Rev.* **73**, 1002.
Wigner, E. P. (1955), *Phys. Rev.* **98**, 145.
Wigner, E. P., and Eisenbud, L. (1947), *Phys. Rev.* **72**, 29.
Williams, J. F. (1975), *J. Phys. B* **8**, 1683.
Williams, J. F. (1976), *J. Phys. B* **9**, 1519.
Williams, J. F., and Willis, B. A. (1974), *J. Phys. B* **7**, L61.
Wladawsky, I. (1973), *J. Chem. Phys.* **58**, 1826.
Wu, T. -Y., and Ohmura, T. (1962), *Quantum Theory of Scattering*, Prentice-Hall, Englewood Cliffs, N. J.
Yarlagadda, B. S., Csanak, Gy., Taylor, H. S., Schneider, B., and Yaris, R. (1973), *Phys. Rev. A* **7**, 146.
Yau, A. W., McEachran, R. P., and Stauffer, A. D. (1978), *J. Phys. B* **11**, 2907.
Ziesel, J. P., Nenner, I., and Schulz, G. J. (1975), *J. Chem. Phys.* **63**, 1943.
Zvijac, D. J., Heller, E. J., and Light, J. C. (1975), *J. Phys. B* **8**, 1016.

Index